Siberia
Conquest and Expansion
(1500-1800)

Siberia
Conquest and Exploration
(1500-1800)

Srikala Balakrishnan

Copyright © 2016 Srikala Balakrishnan. All rights reserved.

No part of this book may be reproduced or transmitted in any form
or by any
means, graphic, electronic, or mechanical, including photocopying,
recording, taping, or
by any information storage retrieval system, without the permission,
in writing,
of the publisher.

ISBN-13:
978-1530637607

ISBN-10:
1530637600

Dedicated
To
The Memory
Of
Satyananda Giri
Who inspired me
By asking
"Why is Russia so enormous?"

Table of Contents

1. The Mongols under Chinghiz
2. The Mongol Invasion of Rus'
3. Stemming the Mongol Tide
4. The Conquest of Kazan
5. The Stroganovs Build an Empire
6. The "Conquest" of the Khanate of Sibir
7. The Siberians
8. Spanning the Breadth of Asia
9. The Amur Valley and Russo-Chinese Relations
10. Finding a Passage between Two Continents
11. The Demidovs and Siberian Iron
12. Bering's First Voyage
13. The Great Northern Expedition
14. The Russian Discovery of America
15. The "Russian Columbus"
16. Siberian Frontier Life
17. Siberian Towns
18. Siberia at the Threshold of the Modern Age

Maps
Illustrations

Chapter-1

The Mongols under Chinghiz

GREAT, CATACLYSMIC EVENTS HAVE THEIR ORIGINS IN SMALL, obscure beginnings. Our story begins in the wilds of Siberia in the Middle Ages. In the year 1177, the Lady Oelun-eke, her three sons and two stepsons were driven into the Siberian wilderness. Banishment into the brutal wilds of Siberia's Trans Baikal lands usually meant death. For a Mongol noblewoman whose husband had just been poisoned by his enemies, the chances of survival seemed even more slim. Once well-fed from Lord Yesugai's numerous flocks and well-tended by many slaves, Oelun-eke and her children now had only the small creatures of the forest, fish caught on hooks bent from needles, and the wild garlic, onions, and roots that grew beside the mountain streams to keep them alive. They were alone. "Apart from our own shadows," she told her children during those dark days," we have no friends."

With the present so fierce and the future so dark, Oelun-eke vowed that her sons must know the glorious saga of their past. Well practiced in the art of "citing ancient words" according to the Mongol chronicles, she told them the story of their origins-of how a great "bluish wolf" had crossed over the waters of Siberia's Lake Baikal to mate with a "fallow doe' in the thick pine forests that clustered beneath the peak of Burkhan-Kaldun. This was the home of Tengri, God of the Eternal Blue Sky and first among the ancient deities of the Mongols. And it was there, at the foot of Tengri's mountain, that the union of doe and wolf had given birth to the first of the line from which the children of Yesugai and Oelun-eke had descended. As the generations passed, Tengri himself had blessed the descendants of this union when he had come to the beautiful Lady Alan after the death of her husband, entering her tent on the shafts of moonbeams and leaving on the rays of the sun after their sons had been conceived. From this union of god and woman had come Kabul Khan, whom the Chinese called King of the Mongols. His son had been Kutula, the Mongol Hercules, with hands like powerful bear claws and a voice that rolled and crashed through the mountain glens like thunder. Yesugai had been Kutula's nephew. Oelun-eke's marriage to him had produced three sons. Temuchin, the eldest had been born in 1167, the Year of the Pig.

Amid the dark forests that spread out between the peak of Burkan Kaldun and the headwaters of Siberia's Onon River, Temuchin grew up, tall and sturdy. He was possessed of an iron constitution, a flint-hard will and an amazing sixth sense that helped him turn the weakness of others to his advantage. Often hungry and without shelter, he became inured to the discomforts of cold, heat, and rain as he struggled to survive in the wilderness. Outcast from the society of the steppe, he nevertheless observed its intricate social hierarchy of lords, knights, commoners and slaves from afar and pondered how they could be made to serve his ends. With that knowledge, he would change the world.

Once he came of age, Temuchin entered the arena of steppe politics. He became the vassal of one of Mongolia's most powerful lords before he was twenty. He built an army of supporters by offering unswerving protection to those who gave him their loyalty. With them he triumphed in great battles and lived through narrow escapes. On one occasion, he languished as a prisoner among his enemies for several weeks before he could break free; on another, only the devotion of a faithful comrade saved him from dying of a wound that had severed a vein in his neck. Always, Temuchin placed his faith in Jebe, Kubulai, Jelme and Subudei, the four great generals who would fight with him throughout his life. Held in check by "chains of iron", these "four dogs" of Temuchin drew joy and sustenance only from the thrill of combat. On the day of battle, the chronicles said, "they eat the flesh of men." At such times, Temuchin soared above them "like a falcon which is become greedy for food."

Temuchin wed the Lady Borte, daughter of an allied chieftain, only to have her abducted by his enemies. Temuchin raised an army of forty thousand and set off in pursuit of Borte's abductors into the thickly wooded lands of the Trans Baikal. There amidst the forests of pine overgrown with rhododendrons and orchids, near what is now Ulan-Ude in the Buryat lands of the present day Russian Federation, Temuchin freed his wife. On the journey home, she gave birth to a son whom she named Jochi, uncertain whether his father was Temuchin or the enemy chieftain who had abducted her. One day, Jochi's son, Batu would conquer Russia.

After Borte's rescue, Temuchin proceeded to weld the warring Mongol tribes into a nation. In the spring of 1206, he summoned "all the people living in felt tents" to a solemn assembly, a Great Kurultai to decide their destiny. These were men and women born to a life of want in the harsh steppe and the bleak deserts of Inner Mongolia. They were ignorant of writing, town life and agriculture. They lived by tending their flock, raiding their neighbors, and worshipping Tengri, God of the Eternal Blue sky. Nature, fate and the predictions of shamans shaped their lives. When their shamans proclaimed that Temuchin had won special favor in Tengri's eyes, the Kurultai proclaimed him Chinghiz Khan.

Chinghiz Khan set out at the age of thirty-nine to build an empire that his sons would extend across three continents. A superbly disciplined, precisely organized cavalry force of more than a hundred thousand men supported by a corps of engineers that had mastered the best siege technologies known in east and West, his army was the most formidable fighting machine of his day. At times advancing in a "lake array" that spread their attack over a large area, at others marching in a "karaquana" formation, so named for the thorny karaquana bushes that grow in thick clumps on the Eurasian steppe, its men lived by the sword and thought only of the present. "The greatest pleasure," Chinghiz Khan once said, "is to vanquish your enemies, chase them before you, rob them of their wealth, see those dear to them bathed in tears, ride their horses[and] clasp to your bosom their wives and daughters.'

Able to travel long distances at great speed, the Mongol fighting men who shared this creed boasted extraordinary powers of endurance. Their razor-sharp broad-headed arrows could inflict massive bleeding and almost instant death upon enemies at close range, while eagle-feathered shafts with lighter heads allowed them to kill at distances approaching four hundred yards. Every warrior carried a quiver of each type of arrow on his saddle and fired them from a large compound bow whose pull of at least 166 pounds far exceeded that of the English longbow. The high front and back of the Mongol cavalryman's heavy wooden saddle allowed him to sit firmly while shooting his arrows toward any point of the compass. In Mongol hands, such weapons had no equals. Not until the development of the breech-loading rifle in the 1860s did the world's fighting men again carry such weapons that enabled them to surpass the accuracy and firepower of the thirteenth-century Mongols. `

A cuirass of thick leather, sometimes reinforced by plates of copper or iron, a helmet of hide and iron that protected head, neck, and throat, a fur or sheepskin coat, a fur hat with ear flaps, felt socks, and heavy leather boots were the Mongol cavalryman's standard equipment for winter fighting. Dried meat, ten pounds of dried curds, a leather bottle filled with two liters of fermented mare's milk, at least two quivers, each with a side pocket to carry a file for sharpening arrowheads, an awl, and a needle and thread rounded out the supplies he carried. If he fought in the heavy cavalry, he carried instead a saber, mace and battle-axe, and a lance that had a cleverly designed hook at the base of its head for pulling enemies from the saddle. All Mongol horsemen led from one to four mounts that survived on whatever tufts of grass they could find along the way. In time of need, their riders could live on mouthfuls of blood sucked from their veins.

Chinghiz Khan organized these horsemen, who could subsist on lice, the blood of their horses, and in moments of real desperation, the flesh of their comrades, into tens, hundreds, thousands and tens of thousands. Each unit shared a ruthless collective responsibility for the actions of its members. If a man fled in battle, the rest of his unit faced death for allowing him to do so. Mongol commanders were expected to order the rest of a unit to be killed if they failed to rescue a comrade who had been taken prisoner. Such a force could be mobilized at short notice, moved toward an objective with great speed, and thrown into battle with devastating effect.

Chinghiz Khan turned these fighting men toward the domains of the emperor of China in 1211. Here was a land where fields of rice, maize and millet had been cultivated for thousands of years by patient farmers skilled in drawing sustenance from the soil. Yet access to this land did not come easily. Wild gorges, gloomy mountain passes, and the Gobi Desert guarded it from the north. Beyond them stood the Great Wall, the barrier that China's rulers had raised a thousand years before to protect their people from the ancestors of the Huns that Attila had led into Europe. Here was a fortress formed by elements that man had tamed and turned to his advantage. To succeed here required more than raw courage; more than even the combined genius of the "four dogs" of Chinghiz Khan.

Not yet experienced in attacking heavily fortified positions, Chinghiz and his generals spent two years forcing a passage across China's Great Wall. Then they turned toward Peking, the capital of the Chinese Emperor and one of the leading metropolises of the world. Surrounded by a triple line of moats, its eighteen miles of stamped clay walls crowned by brick battlements along which twelve gates were interspersed with nine hundred towers, Peking was thought invincible. For two years the armies of Chinghiz stood before it, plotting its destruction and seeking to break its defenses. With the help of captured Chinese engineers, the succeeded in May 1215, after which they carried away a vast hoard of gold, silver, precious stones and rare silks. Legend has it that sixty thousands of Peking's women threw themselves from the walls rather than fall into the hands of Mongol soldiers when the looting began. One eyewitness reported, "the bones of the slaughtered formed whole mountains, and the soil was greasy with human fat." Then the Mongols razed Peking so completely that, Kublai Khan, Chinghiz's grandson had to rebuild it from scratch in order to make it his capital.

Then Chinghiz turned due west. After overrunning the kingdom of Kara-Kitans, which lay to the west of Lake Issyk-Kul, he led his armies into western Turkestan. There stood the flourishing Khwarazm Empire amid a sea of steppe lands and deserts. A powerful Muslim empire turned sour by palace intrigues and heavy taxation, Khwarazm was ripe for attack. Without popular support and with an army whose soldiers preferred to plunder than fight,

Khwarazm's sultan had to face the Mongol cavalry of nearly two hundred thousand men supported by the latest in Chinese siege technology. As the host of Chinghiz Khan entered the Khwarazm lands, it brought with it hundreds of Chinese engines of war that could rain storms of stones and burning naphtha mixed with saltpeter upon a city's defenders from distances well beyond the range of their arrows and javelins.

In the fall of 1219, Chinghiz Khan split his massive force into smaller units and ravaged the countryside. He assembled hordes of captives to storm the better defended cities. In February 1220, these forces reunited at Bukhara. Bukhara was a Central Asian center of learning and trade. It produced plush silken carpets which still bear its name. Bukhara in those days was "the cupola of Islam", a city "embellished with the rarest of high attainments", in the words of the thirteenth century Persian chronicler Ala-ad-Din Ata-Malik Juvaini. Merchants thronged its bazaars. The Muslim faithful flocked to its mosques. But its defenses could not stand the Mongol assault. As they drove native captives against Bukhara's fiercely defended inner citadel, the surrounding moat filled with bodies and gave the Mongol shock troops easy access to the breaches that their siege engines had opened in its walls. "It was a day of horror," an account of how the Mongols sent the survivors into exile concluded. "There was nothing to be heard but the sobbing of men, women and children torn apart forever."

Using captives from Bukhara to spearhead their advance, the Mongols proceeded against Samarkand. A city of half a million, Samarkand was known for its graceful copper urns, fine cottons, silks, paper and exquisite silver lame fabrics. Its canals and fountains had made the desert bloom. Its melons, packed in leaden boxes filled with snow were relished as far away as Baghdad. "If it is said that a paradise is to be seen in this world, then the paradise of this world is Samarkand," one of Khwarazm's poets had written. "A country whose stones are jewels, whose soil is musk and whose rain water is strong wine," another added, Samarkand was the greatest of Khwarazm's treasures. Its defenses were thought to be the strongest in Central Asia. This advantage was however lost the moment the city's commander led its garrison into an ambush that claimed the lives of its defenders in tens of thousands. Attacking three days later to the traditional cacophony of drums, trumpets and bellowing camels, the Mongols overwhelmed the last of Samarkand's wardens. After they sacked the city and leveled its ramparts, the Mongols herded the survivors together to be driven toward their next objective.

Within a year, all of the major cities of Khwarazm had fallen to the Mongols. Chinghiz's youngest son slaughtered nearly three-quarters of a million of the inhabitants of Marv. The same Arab and Persian sources tell us that Nishapur was marked out for special vengeance. The great khan's son-in-law had been killed by an arrow shot from its walls. The Mongols held a blood carnival presided over by the widow in which every captive was decapitated and their heads piled high in pyramids. "Flies and wolves feasted on the breasts of religious dignitaries," the chronicler Juvaini wrote of the fearsome slaughter at Nishapur. "Eagles on mountain tops regaled themselves on the flesh of delicate women," he added, "and vultures banqueted on the throats of houris." Since destruction had been the goal in Nishapur, the Mongols left a fortune in gold, silver, and precious stones among its rubble. In later years, it was said that thirty thousand dinars in treasure could be dug out of its ruins in a single day.

In the weeks that followed, Bamiyan shared the fate of Marv. So did Gurganj where the Mongols destroyed the main dam on the Amu Darya and drowned the entire city. Herat was leveled to the ground after an orgy of killing that lasted a week. At Nisa, near present-day Ashkhabad, the Mongols herded their victims into the fields, ordered them to tie each other's hands behind their backs and then slaughtered them by firing arrows indiscriminately into their midst. The stench of death lay heavily over Turkestan and eastern Persia. Such ruin would take a millennium to repair. "Even though there be generation and increase until the Resurrection," wrote Juvaini forty years later, "the population will not attain a tenth part of what it was before."

While Chinghiz watched the cities of Khwarazm crumble, Jebe and Subudei, two of his "four dogs" of war, rode west in search

Khwarazm's Sultan Mohammad who fled his kingdom in the summer of 1220. In hot pursuit, the two generals chased their quarry into Khwarazm's Persian provinces along the southern shores of the Caspian Sea. The Sultan led them from Nishapur to Sebzevar, and from there to Ray, the greatest city of Persian Iraq. While the Mongols sacked Rey, Mohammad fled west but Jebe and Subudei were at his heels in an instant. Suddenly he turned southwest, perhaps intending to take refuge in Baghdad, and then, just as quickly, doubled back toward the Caspian. On one occasion the Mongol's advance units came close enough to wound his horse with their arrows; on another, he escaped into the Caspian while a hail of Mongol arrows fell around his boat. Finally, at the end of 1220 or the beginning of 1221, Sultan Mohammad died on the Caspian island of Abeskun, leaving Jebe and Subudei free to pursue other ventures.

With twenty thousand horsemen at their command, the two Mongol generals looked for new kingdoms to conquer in the name of their master and the god Tengri. Galloping west across the high steppes that covered most of northwest Persia, they turned north into Azerbaijan, where they rested briefly in the Mugan steppes before they invaded Georgia. In late spring 1221, they swept back into Persia to fight a year-long campaign before they turned north again, forced their way through the Derbent Pass, and broke out onto the Black Sea steppes that make up the present day Kuban. Jebe and Subudei entered the southern lands of Russia where they smashed a small Russian army on the banks of the Kalka River in the southeastern Ukraine in June 1223. Then they crossed the Volga near present-day Volgograd and rejoined Chinghiz's army in Khwarazm.

The Mongols then turned their attention once more to China. The Russians resumed the dynastic feuds that had kept their land in turmoil for the better part of two hundred years. They made no attempt to forge new alliances for defense. Nor did they reconnoiter the lands beyond the Volga to learn where the Mongols had gone. "We do not know where these evil Tartars came from," one chronicler of old Russia confessed. "Whither they went, God only knows." For decade and a half, the fearsome clouds of death that the horsemen of Jebe and Subudei had brought to the edge of Russia slipped back into the horizon.

The passing of Chinghiz Khan in 1227 only spurred the Mongols to greater conquests. Chinghiz Khan bequeathed to each of his sons an *ulus*, that is, a number of tribes and the grazing lands needed to support their herds and flocks. None of the bequests conferred the right to rule the Mongol Empire. The Grand Kurultai bestowed that honor upon Ogodei, Chinghiz's third son. While Ogodei reigned from Karakorum, his brothers and nephews spread their conquests deeper into China, Korea, western Persia, Russia and Eastern Europe. Later, each *ulus* became a powerful kingdom tied to Karakorum only by its loyalty to the great khan. The Persian khanate of Hulagu, the Kingdom of Kara-Kitai, the Mongol Empire of China under Kublai Khan and the golden Horde of Batu all grew out of the conquests that the sons and grandsons of Chinghiz Khan achieved before the middle of the thirteenth century.

Chapter-2

The Mongol Invasion of Rus'

BATU, SON OF JOCHI. SON OF CHINGHIZ, inherited the westward realm in the division of the Mongol Empire. His was *ulus* westward from Siberia's borderlands and the steppes of Kazakhstan "as far as Mongol hooves have beaten the ground." No one knew how far Mongol hooves had actually reached. In 1223, Chinghiz's generals, Jebe and Subudei had crushed a small Russian force on the Kalka river. Now, in the mid-1230s, this expedition west took on a new importance. Batu and his counselors looked west. Of the old guard, Subudei remained alive to tell the tale of the campaign on the Kalka river fifteen years earlier. Batu therefore appointed Subudei, the last of Chinghiz's "four dogs" to be his chief of staff against a campaign west. After crushing the Bulgars near Kazan, Batu and Subudei got ready to attack Russia in the fall of 1237.

Preferring to fight in winter when the great rivers of Siberia and the east froze solid to become smooth highways for the Mongol cavalry, Batu launched his attack. In November 1237, Batu's host of 150,000 cavalry swept across the frozen Volga into Russia. They appeared in front of Ryazan. They offered the people of the city the customary terms of submission-a tenth of all they possessed. Though Ryazan stood close to the Bulgar frontier which had been ravaged earlier the same year, its people suffered from hubris. Excessively complacent, they did not realize the enormity of the danger facing them. They therefore rejected the peace terms. Besides, the prince of Ryazan believed that a stand had to be made lest the principalities of Russia's northeast fall to the Mongols one by one. He and his counselors vowed to halt the Mongol's advance before it got fully underway. Knowing fully well that he did not have the means to oppose the Mongols single-handed, he called on his neighbors for help. This was not forthcoming.

Russia, in those days, was a collection of principalities. Unity was a distant dream. Civil wars were a common occurrence. The domain of every prince was fair game for his neighbor. Always at each other's throats, the princes of Russia paid no heed to Ryazan's call for reinforcements. The death of their countrymen at Mongol hands did not weigh on their consciences. Therefore, Russia's dukes sent no forces to Ryazan to help that city in its war against the Mongols. It was every man for himself.

As the Mongols closed in on Ryazan in December, the people of Ryazan faced the onslaught alone. The brutality with which Batu's generals developed their siege took Ryazan's defenders by surprise. Within five days, Mongol sappers had ringed the town with breastworks and palisades to cut off every avenue of escape. Then engineers attacked Ryazan's walls with ballistae, catapults and armored battering rams. They weakened the defenses and prepared for an assault. On the morning of December 21, Mongol shock troops stormed the city walls. Fierce hand-to-hand fighting followed. Steel clashed against steel. The chronicle tells us, "Men fought with such intensity that even the earth began to moan." After many hours of fighting, the defenders were overwhelmed.

In victory, the Mongols showed no magnanimity. The town became a mound of the dead. "There was not even anyone to mourn the dead," the chronicle of Ryazan's destruction concludes, "Neither father nor mother could mourn their dead children, nor children their fathers or mothers. Nor could a brother mourn the death of his brother, nor relatives their relatives. All were dead." Those whom Batu's archers shot for sport had to consider themselves fortunate. Flaying, impalement and crucifixion confronted the remaining survivors. The list of torments the Mongols inflicted on their victims are endless and horrify the imagination. Suffice to say that they instilled a sense of terror and apprehension on those yet to face their wrath. If they refused the terms offered by Batu's envoys and opposed the Mongols, their failure would prove fatal. Any who took up arms against this terrible foe and failed to triumph would be shown no mercy.

Moscow was the next destination for the Mongols. In political terms this seemed an object hardly worthy of the Mongol's effort. Moscow, was, in those days one of the smaller provincial towns. Its ruler was the lowest ranking of Russia's two dozen or more princes. Yet, Mongol reconnaissance and intelligence thought otherwise. They saw that Moscow stood at the upland in which the great rivers of European Russia had their beginnings. It was the hub from which Russia's river highways zigged and zagged outward like the irregularly shaped spokes of a lopsided wheel. From Moscow, an army could move along Russia's frozen river highways in any direction or in all directions at once.

From Moscow, Batu and Subudei turned to Vladimir. This was the largest city in Russia's northeast. It was the capital of the highest ranking and most powerful of Russia's princes. If the Mongols wanted to advance west, they would have to destroy the army of Vladimir. Vladimir was then ruled by Grand Prince Yuri II. It spread out for more than a mile along the Kliazma river. It was full of onion-domed churches. The white stone cathedral of St. Dmitri was a landmark for travelers for many miles around. The army of Vladimir was better equipped and larger than any in Russia at that time. Yet, it managed to slow the Mongol advance for no more than a few days. Soon Mongol siege engines had breached the city walls and sappers built earthworks and palisades to close a siege round it. On Shrove Tuesday, the Mongols burst into Vladimir and slaughtered the grand prince's mother, wife and children along with every other living soul they could find. The dawn of Ash Wednesday, February 9, 1238, found only the Mongols alive in Vladimir.

Grand Prince Yuri had fled Vladimir before the Mongols closed their ring round the city. He hoped to raise another army to fight the Mongols. As they spread their forces to ravage the countryside and destroy another dozen Russian towns, the Mongols paid him no heed. In the beginning of March, they overran Yuri's badly chosen new positions and slaughtered him and the rest of his men. The Mongols were now free to turn toward Novgorod, the greatest commercial center in Russia. It was the eastern outpost of the Hanseatic League.

There comes a point, often in history, when the tide turns. The tsunami of Mongol terror suddenly halted and spent its force. Too complacent after their easy victories, Batu's force attacked the small outpost of Torzhok. They thought it would to take no more than a day or two to clear this minor obstacle. But Torzhok held out for a full two weeks. By the time the last of its defenders had fallen, it was March 23 and the Spring Equinox. The frozen rivers and lakes of northern Russia began to melt. No longer were the rivers hard cold highways for the Mongol cavalry to ride on. Instead they became death traps ready to swallow man and beast. The Mongols halted sixty miles short of Novgorod and turned south toward the Russian steppe. There they rested for the next two years. Then, in the summer of 1240, Batu again divided his forces. He was now joined by his nephew Prince Mongka who would be the next great khan. Batu and Subudei devastated the towns of southern Russia. They suddenly appeared before Kiev.

Kiev was one of the most beautiful of the cities of eastern Christendom and one of the most important. A hundred churches with gilded domes lay within its walls. St. Vladimir the Blessed who had brought Christianity to Russia in the tenth century was interred in its great cathedral. The remains of the saintly Princess Olga, his mother, and those of Yaroslav the Wise, the eleventh century lawgiver also lay there. During the Middle Ages, when darkness enveloped most of Europe, Kiev remained a center of art and learning, a steppe metropolis. Here the cultural influences of the Middle East integrated with her Slavic past and were seasoned with borrowings from the cultures of western and central Europe. Great brick palaces adorned its center. Libraries, hospitals and schools enriched the lives of its people. This was at a time when London and Paris had yet to come into their own. Kiev was often mentioned in the epics of the Medieval West. Russian folklore speaks of the greatness of Kiev. Its rulers were allied by marriage with nearly every important royal family of Christendom. Apart from Constantinople, there was no city between Baghdad and Vienna to compare with it.

Kiev stood at the crossroads where the caravans carrying spices and silks from the Orient intersected with the traffic that moved along the "Great Amber Road" from the Baltic to Constantinople. Kiev was the "Mother of Russian Cities". It was the bastion that guarded the eastern frontier of Christian Europe against the nomads from the Eurasian steppes for more than two hundred years. Without Kiev, medieval Krakow, Budapest and Vienna could not have flourished. The fierce steppe horsemen who had been driven back from Kiev's walls would have ridden farther west. But by the middle of the thirteenth century Kiev had grown weak. Tarnished by internecine strife, its brilliance had begun to fade. No longer could Kiev's ruler summon the armies of a dozen lesser rulers to his aid. He did not have the wealth to fight large wars as his ancestors had done. Kiev would now present a grand finale to the Mongol ravishment of Russia.

At Kiev, the Russians felt for the first time the full fury of the Mongol onslaught. Prince Mongka showed the ruthlessness necessary to rule an empire. He sent the usual envoys to offer subjugation only to have them killed by a faction in Kiev that was opposed to peace. Now, if the Mongols won, there would be no mercy for man, woman or beast. As at Ryazan or Vladimir, the assault on Kiev would be a fight to the death. Only the greater wealth of Kiev ensured that the battle would be fought on a larger scale.

As the Mongols gathered before Kiev in "thick clouds" in the words of the chronicles, the clamor of creaking carts, bellowing camel and oxen, and neighing horses, all numbering in tens of thousands, aroused a sense of terror and impending doom in the hearts of the defenders. Inside the city walls, people had to shout to make themselves heard. Mongol siege engines closed their ring and started battering Kiev's walls with boulders and rams. Others rained javelins and pots of liquid fire on them. "Arrows darkened the daylight."

After several days of fighting (November 19, or December 6), the Mongols broke into the city. They slaughtered all who remained and ravaged the buildings and churches. The relics of Russia's first saints were scattered to the four winds. The treasures of art and learning that had made Kiev famous were destroyed. "We came across countless skulls and bones of dead men lying about on the ground," wrote the Franciscan friar Giovanni del Plano Carpini after his visit to Kiev six years later. "There are at that time scarcely two hundred houses there." Friar Giovanni found cathedrals in ruins. These stones became foundations for churches built centuries later. More than two hundred years were to pass before a new city rose from the ashes and rubble of the old one. Leaving behind them a trail of devastation the Mongols moved on to Hungary and Poland in 1241. The Russian people became vassals of the Mongols fighting in their armies as far west as Hungary and as far east as China.

Russia then, was not yet Russia. Torn by civil war, the princes of Russia had no sense of national destiny. The unity of Russia was a distant dream. Brother fought against brother, cousin against cousin and neighbor against neighbor. Paying tribute to the Mongol was, in their eyes, no worse than paying tribute to a stronger prince. It was a simple matter for the Mongols to hold them in servitude. Batu had come and gone and the princes of Russia continued their civil wars. Until they developed a sense of national identity and united, they could never hope to shake free the Mongol yoke.

Eventually, one prince was to stand out and shine in this firmament of Russian grand dukes. He was the grand duke of Moscow. The princes of Moscow forged Russia into a single nation from a bevy of duchies and small principalities. Overcoming their rivals, sometimes by diplomacy, at others by treachery and resorting to arms only when all else failed, the princes of Moscow added Vladimir, Suzdal, Rostov and a score of other principalities to their domains. In the process, they gained new resources of men and treasure. This was the historic "gathering of the Russian lands" that Grand Prince Ivan I began in the 1320s. Six generations later the gathering would be complete. By then, Moscow would become Russia.

Ivan I was a schemer among schemers. Karl Marx described him as a ruler who combined the "qualities of a Mongol hangman, a sycophant and a slave-in-chief." but his people called him Ivan Kalita, i.e., "Ivan the Moneybag." He earned this nickname from his well-known reluctance to part with Moscow's treasure. He was careful to curry favor with the Mongols who had by then built a capital called Sarai near present day Volgograd. Ivan Kalita became one of their tax collectors. A part of what he took in the name of the khan remained in Moscow. With this new treasure he gained the upper hand among his feuding rivals.

The princes of Moscow were by no means the only of Russia's many potentates who courted the Mongols at Sarai. What distinguished Ivan and his descendants was their strong adherence to the Russian Orthodox church. The metropolitan of this church became Moscow's chief ally. While their enemies fell by the way, the princes of Moscow gained by their wealth, political power and church support. They became the successors of the Mongols as masters of the Russian land.

The monkish chroniclers who recorded the gathering of the Russian lands described the rulers of Moscow as the earthly agents of Christ. Pious churchmen proclaimed that any who followed them were doing the work of God. Moscow's rulers, so the church claimed, had the mandate from heaven to unite the Russian land and free it from the Mongol yoke. Combined with rivalries among the descendants of Batu Khan that produced erratic and muddles policies towards Russia's princes, the urging of the Church fathers convinced Grand Prince Dmitri to test the strength of the Mongols. This he did in the memorable battle of Kulikovo Field fought on the banks of the Don, on the Feast Day of the Nativity of the Virgin, September 8, 1380.

Chapter-3

Stemming the Mongol Tide

THE ISSUE BETWEEN THE MONGOLS AND THE RUSSIANS was decided in an engagement on Friday, September 8, 1380. This day was witness to a conflict between the two armies on Kulikovo Field on the banks of the River Don. A large Mongol force of Khan Mamai was marching in the dense fog to punish Moscow's grand prince for raiding one of their expeditions two years before. This was the largest Mongol force to invade Russia since the days of Batu. It was expected to be even more effective than the Steppe horsemen Batu had led. It was strengthened by elite infantry units hired from the Genoese settlements in the Crimea which were to anchor the center. This would leave the Mongol cavalry free to fight in both flanks. Besides, the armies of Gran Duke Jagiello of Lithuania were expected to join them in a few days' time. Together, the Mongols and the Genoese numbered thirty thousand; Jagiello's force would increase their ranks by half that amount again.

The union of two such forces seemed to promise to give Mamai an enormous advantage over the Russians. Certain of that and confident that the Russians were still far way, he called in the shield of cavalry that usually served as the eyes and ears of advancing Mongol armies. Now, he and his men were waiting in Kulikovo Field, the "field of snipes" until Lithuania's grand duke joined them. Once united, they planned to march on Moscow.

While the Mongols and their allies assembled on Kulikovo Field, an army of Russians was marching towards the Don. Like the Mongols, they numbered thirty thousand. From spies and informers, they knew the Mongol's plans for a joint campaign with the Lithuanians. They also knew that both forces thought they were still in Moscow. Hoping to take the Mongols by surprise and defeat them before the two armies united, Prince Dmitri of Moscow advanced quickly. He sent horsemen to reconnoiter the land ahead of his main force. He had already chosen as his battlefield the same Kulikovo Field where the Mongols were waiting for the Lithuanians.

Kulikovo was a slightly rounded spit of land formed by the junction of the Don and the Nepriadva. It seemed ideally suited for Dmitri's purpose. The deep streams and thickets that cut across it would prevent the Mongols from using their cavalry to sweep up his flanks once battle was joined. The larger rivers on either side would prevent the Mongols from attacking his rear. From Dmitri's perspective the only dangerous terrain lay to his right. If the Mongol cavalry won enough room to force his army against the steep banks of the Nepriadva, it would be hard pressed to escape. A shallow ford across the Don left the way open for retreat.

On that fateful morning, the Mongols were gathered round large kettles of food after marching for several days. In the dense Don fog, nothing was visible. Dmitri moved his men, imperceptibly into position. Certain of salvation in the next world if not in this one, the Russians looked forward to a "second baptism" as they fought under the black banner of their prince. "Let us place upon our heads the crown of victory!" Dmitri cried as he donned the armor of an ordinary Russian soldier. "If I die, it will be among you," he told the men who would be fighting to his left and right, "If I survive, it will be while fighting among you also." As the fog lifted, the Mongols saw the Moscow men advancing on them. As they seized their weapons, they did not see the reserves of cavalry that Dmitri placed in ambush in the large thicket that stood on the Russian's left.

The Battle of Kulikovo Field began at about eleven in the morning. Each side sent forth a champion to fight in single combat. The men struck each other with their lances simultaneously. Both died the same instant. Then the Mongol and Russian armies fell on each other. Russian straight swords clashed with Mongol scimitars. Neither side gave ground or gained it. Blood, the chronicle says, "flowed as in a cloudburst". Some of Prince Dmitri's men saw his horse fall from under him round midday. Others saw him remount. Still others saw him fight on foot after that limping and wounded. He was still in battle when the balance began to tilt towards the Mongols midafternoon. Then, just as the khan's armies were about to claim victory, the reserves Dmitri had hidden in ambush fell upon their rear and drove them from the field. Not until evening did the Russians find their prince, unconscious yet living, beneath a pile of corpses. Twelve Russian princes, five hundred great lords and almost half the army that had fought on the Don that day perished.

It took eight days for the Russians to bury their comrades who had fallen that day in a common grave in Kulikovo Field. They left laden with trophies. Dmitri proclaimed "Donskoi" by his people had helped turn the tide against the Mongols. He was to die before the age of forty from injuries he had suffered during battle. Still another century would pass before the Russian's final liberation from the Mongol yoke. This could happen only after the Russians forgot their internal conflicts and unified under one sovereign.

Although a historic moment in Russia's history, Dmitri's victory did nothing to wipe out the political instability which undermined Russia's unity. Even though fear of Mamai's invasion had prompted some of the Russians to take up arms against him, others had sided with the Mongols all along. Four great princes including the rulers of Novgorod and Tver had sent no troops to Dmitri's army in 1380. The victory over Mamai's host had divided even those who had. When the Mongols attacked Moscow two years later no more than a handful of allies helped Dmitri defend his capital. Half a century after Ivan Kalita had begun to gather the Russian lands, the Russians seemed more divided than ever.

At the same time, the Mongols were divided by internecine quarrels. Ever since Janibeg Khan had been murdered by his son in 1357, Mongols had been fighting Mongols throughout the lands of the Golden Horde. Mamai was only one among several khans who rose to rule over Batu's realm in the 1370s. Assassination to eliminate a rival khan was a known technique among the Mongols. At one point, five khans were murdered in five years. As Mamai fought the Russians at Kulikovo, an explosive situation developed among his own troops. Even as forces obstructed the unification of Russia, still other forces pulled the Mongols apart. A strong man was needed to unite the Mongols even as the Russians needed a strong leader to unite them.

Such a hero rose in Central Asia even as Dmitri fought his first wars in Russia's northeast. Born in the city of Kesh in 1336, the Year of the Mouse, Timur the Lame was the leader to revive Mongol fortunes. A brilliant soldier, a patron of art, literature and the law, and a prudent leader who viewed the world through Machiavellian eyes, Tamerlane became a practitioner of *realpolitik*. He was the builder of the last of the great Mongol empires in Eurasia.

Tamerlane soon conquered much of the old Khwarazm empire. He built a new capital at Samarkand. This city, rising out of the ashes of the devastation caused by Chinghiz's armies, was adorned by a dazzling array of splendid buildings. It was to be the last of the great Mongol cities. Here Timur built the Mosque of Bibi Khanum in honor of his favorite wife. Its main portal was flanked by minarets overlaid with exquisite mosaic tiles and its courtyard was dominated by a huge Koran desk formed from two colossal wedge-shaped stones. There were mausoleums for each of Tamerlane's sisters. On the other side of the citadel was placed Timur's own tomb, the Gur Emir whose massive sky blue dome still rises above his city.

In the late 1370s, Tamerlane struck an alliance with Prince Tokhtamysh, a disgruntled nephew of the khan of Kazakhstan. Tokhtamysh was a brilliant general who needed the smallest amount of encouragement to march against the fragmented lands of the Golden Horde. Quickly, Tokhtamysh conquered the eastern parts. As Mamai advanced against Dmitri at Kulikovo, he attacked Mamai's rear. Mongol was pitted against Mongol. Not far from where Jebe and Subudei defeated the Russians in 1223, Tokhtamysh shattered Mamai's army in 1381. Then, he, not the less able Mamai took the field against the Russians.

Supported by Dmitri's Russian rivals, Tokhtamysh marched on Moscow rebuilt after Batu's invasion, in brick and stone. After fighting for three days and three nights without success, Tokhtamysh resorted to treachery. He used a flag of truce and false promise of negotiations to induce Moscow's defenders to open its gates. Then his army fell upon the defenders. They slaughtered them in the streets, in their homes and in the great stone churches where they sought refuge from the flames that engulfed their city. "All at once her beauty perished and her glory disappeared," the chroniclers of Voskresensk wrote of Moscow after the Mongols had finished their work, "Nothing could be seen but smoking ruins and bare earth and heaps of corpses." Having gone north to raise reinforcements before Tokhtamysh began his attack, Dmitri returned to find a capital in ruins and more than twenty thousand corpses. Russia's princes started hurrying to pay tribute to the Golden Horde's new ruler. Unable to unite his allies of the Kulikovo campaign, Dmitri submitted as well.

Dmitri renewed his pledge to the Golden Horde at the end of 1382. The gathering of Russia's lands seemed to have fallen on bad days. However, Tokhtamysh and Tamerlane were soon at each other's throats. Throughout the 1380s and 1390s, the two Mongol princes fought each other in wars that raged across Dagestan, Transoxiana and Kazakhstan. Tokhtamysh was left with only a fragment of his once mighty host. Yet fate kept them from a final confrontation which would have meant the death of one or the other. Tamerlane died in 1405 as he set out to conquer China. Tokhtamysh, who had found refuge in the remote khanate of the Tatars of western Siberia, died at about the same time even as he was assembling one more army to hurl against his formidable rival. After that, the Mongol's power declined. The Russians were finally welded into a nation by Dmitri's grandson Vasili the Blind.

The final confrontation between the Russians and the Mongols took place in October 1480, a century after Mamai's defeat. Once again the khan began his war by making an alliance with Lithuania, Russia's greatest enemy in eastern Europe. The Lithuanians failed him just as they had failed Mamai. This time, however, unlike in Kulikovo, there was no pitched battle between Russian and Mongol. After several months of facing each other from positions across the river Ugra, the Mongols retreated into the Steppe. The Grand Prince of Moscow replaced the khan as the supreme authority in Russian lands.

At his death in 1505, Ivan the Great, who had led the Russians in the final conflict against the Mongols, had almost united the Russian lands. A few finishing touches in the north and east were left to his son. Moscow's grand princes were to become the tsars of all the Russias. They were the overlords of all lands west of the Urals. However, Moscow's Grand Prince or the Tsar had to contend with formidable external rivals. To the north, Sweden barred Russia's way to the Baltic. Poland joined by marriage ties with Lithuania became the most powerful state between Russia and the Holy Roman Empire.

With the road to conquest closed in the west, the Russians turned east. In the second half of the sixteenth century, the Russians began to look to the treasure filled fur lands of Siberia. Ivan the Great sent several punitive expeditions against the natives of Siberia's northwest during the last years of his reign. One of these expeditions captured forty native villages and a thousand prisoners. Yet to approach Siberia from the northwest was too difficult and arduous. Some five hundred miles to the south lay an easier route through the Tatar khanate of Kazan. In the days of Ivan the Great, it was still independent. To claim Siberia, Kazan would have to be taken.

Chapter- 4

The Conquest of Kazan

IT WAS LEFT TO IVAN THE TERRIBLE, GRANDSON OF IVAN THE GREAT to accomplish this task and open the way to Siberia through Kazan. Tall and strongly built, with small piercing blue eyes, Ivan shaved his head and wore his beard thick in the true Muscovite fashion. Daring, crafty and cruel, he inspired fear in a greater measure than any of his predecessors. A recent examination of his skeleton by medical experts has revealed that he suffered from a crippling spinal disease in his later years. We do not know if it hindered his efforts to rule Russia. Nor do we know the extent to which the drugs and alcohol he consumed to deaden his pain affected his judgment. The sources for Ivan's reign are still murky. We still do not know if he wrote some of the epistles attributed to him. Some experts are not sure if Ivan knew how to read or write.

Ivan began to rule in his own right in 1547. The remains of three hundred years of Mongol domination were still extant. The grandsons of the men who had opposed his grandfather in 1480 ruled the broken fragments of the empire of the Golden Horde. The Tatars, a fringe element in Batu's time, came to the fore as the Mongol empire fell apart. A number of small Tatar states were scattered through western Siberia and Kazakhstan. None of these states seriously challenged the tsar's authority in Moscow. However, allied with the Vogul and Ostyak tribesmen in Siberia's northwest, they proved a nuisance. Ivan the Terrible made the stronghold of Kazan his first foreign venture because the Tatars of Kazan were the most bothersome of all.

Fortified by the prayers of the Church and with the certain knowledge that he had been called upon by God to punish the treacherous Tatars, Ivan resolved to conquer Kazan with a deep awareness of his responsibility as the divinely appointed ruler of Russia. First it was necessary to ensure that divine influences would work in his favor. He set out on the inevitable pilgrimage to the Troitsa-Sergeyevsky Monastery. After praying there and at the Uspensky Cathedral in the Kremlin followed by prayers at the tomb of St. Peter the Metropolitan, Ivan felt sufficiently blessed to undertake the Kazan enterprise.

At the head of a formidable army, drums rolling and banners waving, Ivan set out for Kazan early in the morning of Thursday, June 16, 1552, leaving his wife Anastasia, Makary and his brother Yury to govern his country. While he was on the march, a messenger from Putivl brought news that the Crimean Tatars had crossed the border. He ordered the army to march to the heavily defended fortress city of Kolomna. There, the army waited for the Crimean army to attack. For a few days nothing more was heard of the Crimean advance. Then there came a report that about seven thousand Tatars had appeared before the gates of Tula only to vanish again. On the evening of June 23, Ivan learned that Devlet Guirey, Khan of Crimea, had reached Tula with his vast army composed of Tatars and Janissaries, and with heavy Turkish cannons. He gave orders for his whole army to march on Tula, a distance of about 75 miles.

The siege of Tula was soon over. The defenders, emboldened by news that the Tsar's army was on the way, fought brilliantly. The Khan lifted the siege and ordered a retreat. Ivan was overjoyed. He returned to Kolomna to offer thanks to the Virgin in the cathedral. With this defeat of the Khan of the Crimea, the way was now open for the march on Kazan.

The plan of the campaign was carefully worked out. From Kolomna the army would march to Sviazhsk in two columns, the northern column taking the road through Vladimir and Murom, the southern column taking the road through Ryazan and Meshchera. The two columns would meet at a crossing of the Sura River before advancing on Sviazhsk. The northern column was commanded by Ivan and the southern by Prince Ivan Mstislavsky and Prince Mikhail

Vorotynsky. On July 3, 1552, after once more praying before the Icon of the Virgin in the Cathedral at Kolomna, Ivan gave the order to march. In five days, the main army reached Vladimir. Here following his custom, he prayed before the relics of the saints and invoked on his army the blessings of God and the Virgin. He spent a week in Vladimir, and on July 13 reached Murom. Here he received a long and strenuous letter from the Metropolitan Makary, at once blessing and rebuking him. On March 20, Ivan marched out of Murom with his troops and two weeks later reached the Sura River which marked the boundary of Russia and Tatary. On the same day by prearrangement, the southern column also crossed the river. Prince Andrey Kurbsky, who rode with his column, remembered the journey to the Sura River with horror, for their food soon ran out. While riding across the uninhabited Steppe they lived on fish and whatever wild animals they could catch.

The worst part of the march was now over. The great cavalcade came in sight of Viazhsk. It was greeted by the townspeople overjoyed at the safe arrival of the Tsar. In Sviazhsk, the Tsar set up his tent in a meadow outside the walls. In this sumptuous tent, he held another council of war attended by Vladimir of Staritsa, the boyars and the voyevodas. Khan Shigaley also attended the conference. He was instructed to write a letter to Khan Yediger Makhmet demanding his submission. If he came to the Tsar, he would have nothing to fear and would receive many rewards from the Tsar's hands. At the same time, the Tsar sent another letter addressed to the chief mullah and all the Tatars living in Kazan, promising that if they submitted all their past acts of rebellion would be forgiven. These letters were sent on August 15, two days after the army reached Sviazhsk.

Ivan had no real hope that Khan Yediger Makhmet would surrender the city. The letters therefore were merely formal statements of his claim to Kazan.

Without waiting for a reply, he ordered the army to begin crossing the Volga the next day. Two days later, on August 18, he crossed the river with his bodyguard. It was not until the following day that the whole army reached the left bank of the Volga. In darkness and rain, the army drove toward Kazan, the wheels of the carts and carriages clogged with mud. The heavy guns, which came down by boat, were landed at a point four miles from Kazan. The chief obstacle was the small and swift-flowing Kazanka River, but it presented little difficulty. Six bridges were thrown over it. By August 20, the whole army had crossed over.

From his camp at the mouth of the Kazanka River, the Tsar saw the fortress city of Kazan. The city followed the pattern of many ancient and medieval cities: there was the acropolis, the fortress on the heights, while below it, sprawling across the plain, lay the lower town with its huddled streets and long avenues leading to heavily fortified gates. Here and there the lower town was cut by ravines, and sometimes the huddled streets opened into gardens and lakes. In the lower town lived the merchants, the artificers, the workmen, the poor, and the soldiers who manned the wooden towers.

To defend the city, Khan Yediger Makhmet had an army of about 30,000 well-trained soldiers and about 2,700 Nogay tribesmen. They were armed with bows and arrows, spears, swords, lances, maces and muskets. They also had heavy guns and ample supplies of gunpowder, and there was enough food in the city to take them through a long siege. Both the Tatars and the Russians wore chain mail and pointed iron helmets, so that it was sometimes difficult to tell them apart.

Ivan was under no illusions about the dangers confronting him. Khan Yediger Makhmet was a determined, daring and ruthless opponent. He could be expected to use every ruse to prevent the city from falling into Russian hands. There was no possibility of a sudden surprise attack. It was not simply a question of conquering a well-defended city. There was also the problem of how to deal with the Tatar armies outside the city, the many soldiers based in the town of Arsk, which lay on the other side of a dense forest stretching almost to the walls of Kazan. All together there were about 35,000 Tatar and Cheremiss troops loyal to the Khan outside the city, most of them hidden in the forest of Arsk. Of necessity, Ivan would have to take special measures to protect his rear, his lines of communication with the supply ships moored on the Volga, and his own person, for the Tatars well knew that if the Tsar was killed or captured the siege would be lifted.

The plan of campaign was carefully worked out by the Tsar's war council. Each army was led by two generals, one senior and one junior. Thus the main army was led by Prince Ivan Mstislavsky with Prince Mikhail Vorotynsky acting as his second-in-command. His brother, Prince Vladimir Vorotynsky, commanded the Tsar's elite troops, with Ivan Sheremetev the Elder acting as his second-in-command. In addition to the main army and the elite corps, there were seven other armies. They were called the vanguard, the rear guard, the right wing, the left wing, the scouts, and the armies of Vladimir of Staritsa and Khan Shigaley. Each army was given its own separate task, and all armies were under the command of Prince Ivan Mstislavsky.

By August 20, this entire army with its heavy cannon and war machines was standing on the shores of the Volga waiting for the order to storm Kazan. But the order did not come. Prince Ivan Mstislavsky and the council of generals decided to move cautiously, wanting to learn more about conditions inside the city before attacking. On that day, two persons entered Ivan's camp, each of them bringing information of the utmost importance. One was a Russian, a former prisoner of the Tatars, who was allowed to go free on the condition that he give a letter written by Khan Yediger Makhmet to Khan Shigaley. This letter denounced Khan Shigaley as a scoundrel and traitor and went on to denounce the Tsar and the Orthodox Church with extraordinary savagery.

The letter from Khan Yediger Makhmet gave some indication of the temper of the defenders. Kamay Mirza was the second person to escape from Kazan. He was a Tatar nobleman who slipped out of the city with seven companions, the survivors of about two hundred men who had hoped to join the Tsar's forces but had been arrested and executed. He brought news about the city's defenses. He reported that the Tatars were well-armed and well-equipped, numbered about thirty thousand, and there was another Tatar army under Prince Yepancha hiding in the forest of Arsk. This was alarming news. During the two following days the council of generals decided upon the final disposition of the troops outside the walls of Kazan. The order to march was given during the early hours of the morning on August 23.

The plan of the campaign was based on the assumption that Kazan could be conquered only after a lengthy siege. The main army would be stationed outside the east and south walls, the vanguard outside the north wall, the rear guard and left wing outside the west wall, and the scouts in the marshy ground south of the Kazanka River facing the acropolis in the east. The brunt of the fighting would fall on the main army, which would also have to fight the Tatar troops coming from the forest of Arsk. Since the walls of Kazan were over twenty-four feet thick, they could be breached only by blowing them up with gunpowder. All round the city the soldiers were ordered to build earthworks, which took the form of enormous wicker baskets about eight feet high and seven feet in diameter, known as gabions. They were solidly packed with earth to protect the guns and to provide defense works against the Tatars issuing out of the gates in sudden attacks against the Russian troops.

When the soldiers looked up at the walls of the city, they saw no sign of life. No guards manned the towers, the gates were closed, and the city seemed silent and deserted. Many Russians rejoiced, imagining that the Tatars had been overcome by fear and had fled to the forest. Others, who knew the enemy better, advised caution. The Bulak River, little more than a muddy stream, followed the west wall of Kazan, and beyond the river and a small ridge lay the plain of Arsk. Suddenly, as an advance patrol of about seven thousand scouts marched over the ridge, the great Nogay Gate flew open and streams of Tatars poured out of the city to attack them. The Russians were taken by surprise. About five thousand Tatar cavalry armed with lances hurled themselves on the scouts. Another thousand Tatar bowmen came running out of the gates. The scouts were forced back over the ridge and they would have been cut down to the last man if the vanguard under Prince Ivan Turantay-Pronsky had not rescued them. Finally, the Tatars were driven back to the gates. They fought well; they had achieved surprise, and only ten prisoners had fallen into Russian hands, but they had lost the skirmish. The Tsar was well pleased with his small victory, but he knew that at any moment the gates would open again and the Tatars would come streaming out.

On the night of August 24, there rose a terrible storm. Many Russian ships were sunk and vast supplies of food and ammunition were lost. The Khan's Meadow became a shallow lake. The storm lasted throughout the night and subsided early in the morning. On August 25, the Tsar was seen riding around the walls of the city and inspecting his troops. The heavy gabions were rolled into place during the day and the following night, and by morning Kazan was surrounded by a ring of earthworks. Then the heavy guns were brought up and placed behind the gabions.

With Kazan enclosed in a wall of earth, it soon became clear that Prince Yepancha would attempt to break out of the forest of Arsk. Although the Russians knew about this force, they had not expected it to emerge for some time and had merely stationed a few detachments of cavalry along the edge of the forest. The first sortie from the forest took place on August 28, a Sunday. The Tatars came surging out of the forest taking the Russians by surprise. They killed the commander of the cavalry detachment. The whole cavalry detachment might have been killed if reinforcements from the vanguard and the main army had not arrived in time. This first battle on the Plain of Arsk taught the Russians a lesson they would not forget: there was no safety as long as Prince Yepancha remained in the forest. Attacks from the forest were concerted with sudden sorties from the city gates. Occasionally too, the Cheremiss tribesmen attacked the Russians from the northwest pouring out of another forest. They were ill-equipped and no match for their enemies, but they tried men's patience. The Tatars on the acropolis signaled to their friends on the edge of the forest of Arsk by means of battle flags flying from the huge tower dominating the city. The Muscovites also had a carefully worked out signaling system using heavy drums.

Already Ivan was coming to the conclusion the siege might last through the winter. He had no doubt Kazan would be conquered. His hopes rested on God, his army and his sappers. Through Kamay Mirza the war council learned that the main source of Kazan's water supply was a hidden spring which rose on the banks of the Kazanka River and fed into the city through a secret underground passage. Since the spring lay outside the walls, the Russians could blow up the passage if they could find it, and the people of Kazan would die of thirst. There were no springs inside the city, only some brackish pools and lakes.

A stone bathhouse on the northwest of Kazan had already been captured. It was logical to assume that the secret passage passed near the bathhouse. A certain Razmysl, an engineer of Lithuanian origin, took charge of the operation, and Alexey Adashev, the Tsar's favorite was given overall command, thus emphasizing the importance of a ruse that might bring the siege to a quick end. The sappers started digging on August 26 and in ten days reached a point underneath the secret passage. Eleven barrels of gunpowder were rolled into the tunnel. At dawn on Sunday September 4, the gunpowder was ignited. Huge logs, stones and rubble were hurled into the air; the wall caught fire; many Tatars were killed. The people of Kazan were dumbfounded by the destruction of their water supply, and many began to talk of surrender. They dug into the rock for another source of water. They found only a small spring so brackish it was scarcely drinkable. Some people became ill and swollen drinking from it; some died. Nevertheless, the Tatars went on fighting.

The sappers continued to dig tunnels under the walls of Kazan. Two fortified towers, one on the southwest corner and the other on the eastern wall were both mined. The work proceeded slowly and was not completed till the end of September. At the same time the Russians bombarded the Arsk gate with their heavy guns, finally destroying it. However, the Tatars put up a new wooden gate very quickly. Meanwhile mortars kept lobbing stone balls over the walls day and night.

Prince Yepancha, from his hiding place in the Arsk forest, continually made sorties against the Russians and this incessant harassment grew costly. Finally, the Russians lured him into the open plains and defeated him. The remnants of his force took refuge in a fortress built of huge balks of timber, situated on a hill surrounded by swamps. At a council of war, the decision was made to destroy this formidable stronghold, whatever the cost. Prince Alexander Gorbaty-Shuisky was entrusted with the command of the punitive expedition; he was ordered to destroy the fortress and capture as many prisoners as possible. If the fortress was destroyed within a short time, his next task was to advance on Arsk, situate on the Kazanka River some twenty miles beyond.

Prince Gorbaty-Shuisky rode into the forest with his cavalry. When the Russians reached the swamps at the foot of the hill, the cavalry dismounted and spilt into two groups, one group mounting a frontal assault on the fortress with bowmen and musketeers, while the other, led by the prince, made a surprise attack on the right of the fortress after hacking their way through dense forests. The battle was all over in two hours, the Tatars fleeing with the Russians in pursuit. About a hundred Tatars were captured and a vast amount of booty. Two days later, the Russians reached Arsk to find the town deserted. They came across estates of the Tatar nobles, rich fields, cattle, and grain, stores of honey, and many villages which the Russians raided at their pleasure. On these estates they found Russians working as slaves. They were now liberated and joined the Muscovites in marauding expeditions that led them to the banks of the Kama River. The army finally returned to Kazan, bringing not only the liberated Russians and a multitude of prisoners but also herds of cattle and wagonloads of fur and treasure.

The siege was going well, though outwardly there was little to show for the constant hammering on the walls of Kazan. Ivan was waiting for the moment when Razmysl would report to him that he was ready to blow up the two towers. Through these breaches his army would pour into the city. A powerful siege engine was being built secretly some distance south of Kazan. It took the form of an enormous wooden tower forty-two feet high, considerably higher than the walls of the city. The formidable armaments, arranged on the top stages of the tower, consisted of ten heavy guns and fifty light cannon, the heavy guns being ten feet long and the cannon seven feet. They were manned by the best gunners in the army.

The tower which took two weeks to build, was rolled up to the

Khan's Gate during the night. At dawn there was a thunderous roar as the guns fired directly into the city causing fearful damage and killing vast numbers of women and children. The Tatar soldiers behind the Khan's Gate quickly dug trenches or put up earthworks but the presence of the huge tower bristling with guns was a constant reminder of the massive power of the invaders.

As they saw themselves more and more tightly encircled, the Tatars made more sorties, fighting desperately at the gates, hoping to bring about such heavy losses that the Russians would raise the siege. Everything now depended on the sappers, who had been digging for a whole month. The quiet work under the earth was completed on Saturday October 1, the Feast Day of the Intercession of the Virgin Mary. On that day, Razmysl reported that all tunnels had been completed, the gunpowder barrels were in place, and it remained only to light the fuse. The war council decided that the city should be stormed at dawn the next day.

On Saturday the final preparations were made. There was an especially heavy bombardment as though to prepare the Tatars for what was to come. Wherever possible, the moat around the city was filled with earth and tree trunks, so that the Russians could break through wherever the walls showed signs of weakness. On the morning of October 2, the two towers were blown up. The general assault began on all sides of the city, but especially along the east and south walls, where the Russians hurled themselves through the breaches. At the same time, scores of Russians were slipping away from the city not because they were being vanquished by the enemy but because they wanted to safeguard their loot and bring it back to their camps. The Tsar's military advisers ordered the elite corps of mounted cavalry, all of them noblemen, to be thrown into the battle, to fight the enemy and also put an end to the looting.

Prince Kurbsky was not present at the Khan's Gate. With his brother Roman, he was fighting strenuously along the north wall of the city, attempting to capture the acropolis. The Tatars fled beyond the Kazanka River, and while Prince Kurbsky was charging them with three hundred of his cavalry, he fell from his horse and would have been trampled to death if he had not been wearing heavy armor.

'I had so many grievous wounds that I lost consciousness," he wrote. "When I came to about an hour later, I saw two servants of mine and two soldiers of the Tsar standing over me, weeping and sobbing as though I were dead. And I saw myself lying naked, wounded in many places but still alive, for I had been wearing a very strong armor inherited from my forefathers."

By this time the fighting was nearly over. Khan Yediger Makhmet retreated to his walled palace on the acropolis and continued to fight off invaders. At last they broke into the palace, where they slaughtered men, women and children indiscriminately, until the whole acropolis was running with blood. Kul Sherif, the chief mullah of Kazan led a desperate charge against the Russians, but the mullah and his men were slaughtered. Everywhere the dead lay in heaps. They lay in narrow alleyways of the city, in the palaces, in the mosques, and they were piled high against the walls. The Russian chroniclers speak of the whole Plain of Arsk being carpeted with bodies, suggesting that the Tatars made massive sorties before they were cut down.

Escaping from his palace, Khan Yediger Makhmet took refuge in one of the fortified towers still remaining in Tatar hands. Here Prince Dmitry Paletsky parleyed with him, urging that he surrender because his cause was lost and it was time to put an end to the fighting. From the tower, the Khan saw a city in flames and the Russians in full possession of the acropolis. He offered to surrender, but the remnant of his army, observing the fate of their countrymen, decided on flight. While, the Khan, his wife and his court went into captivity, he remaining Tatar soldiers succeeded in scrambling over the walls in the hope of making for the forests beyond the Kazanka River. Most of them were killed by the Russian cavalry.
Elsewhere in Kazan, all resistance came to an end. There was only the wailing of the women and the crackling of the fires. The fighting, which began a dawn, was over by early afternoon.

The ceremonies of victory were performed in the manner of medieval romances. The Tsar remained on horseback while all his generals bowed low before him. Khan Yediger Makhmet, his wife and courtiers were brought before him in chains. He pardoned the

Khan. It was a moment of exquisite triumph never to be repeated.

The victory over Kazan had momentous consequences. The Russians had overcome the last obstacle on the path to the conquest of Siberia. They won for Russia control of the river routes that flowed from the crest of the Urals to the Volga. The way to Siberia was now open. Daring entrepreneurs made ready to take advantage of the tsar's conquest. At their head stood Anika Stroganov. This man's uncanny business instincts and sense of timing and opportunity were to make him the richest man in Russia. Using the Kazan gateway, he and his sons were to build a private empire on the edge of Siberia. They would reach high and become one of Russia's noble families.

The Russian's conquest of Siberia was quite different from the Mongol's invasion of Russia. It exhibited none of the brutality of the Mongols nor did it inspire any terror. It was marked by a flurry of skirmishes, a few small battles, and a handful of frontier forts and trading posts scattered along Siberia's westernmost rivers in the 1580s and 1590s. It is not even certain if it could be called a "conquest" nor when it occurred.

Chapter-5

The Stroganovs Build an Empire

LEGEND HAS IT that the name Stroganov is derived from the fate of Spiridon, an aristocratic Mongol ancestor who fled the service of the khan. Spiridon joined the entourage of Dmitri Donskoi and converted to Christianity. He served the Russians and married one of them, earning a reputation for loyalty. To the Mongols, however, he was a traitor. During Tokhtamysh's war against Moscow they arrested him and executed him by peeling his flesh off in strips, layer by layer, until nothing but his bones remained. Spiridon's widow fled, finding refuge in the valley of the Dvina river in Russia's remote northeast. Here, she and the posthumous son she bore Spiridon were safe, with hundreds of miles of virgin forest between them and the vengeful Mongols. She commemorated her husband's martyrdom by taking the name Stroganov (from the Russian verb, *strogat,* meaning, "to peel"). She founded the family who became keepers of Siberia's western gateway.

The Stroganovs became traders in Solvychegodsk a town which sprang up on the banks of a newly discovered salt lake on Russia's northeast frontier. There, Luka, Spiridon's grandson started dipping out water from the lake and boiling it in large rectangular iron pans. He collected the fine residue that remained after the liquid evaporated. Salt, in those days, in Russia, was nearly as valuable as gold. It was the only way of preserving food. Without native sources of salt, the Russians were obliged to depend, for their supply on North Europe's Hanseatic League and on the Genoese who colonized the Crimea. They paid dearly for life's bare necessity.

Now the Solvychegodsk salt works offered a chance to end this dependence on foreign trade. Eventually, the industry expanded to produce huge quantities of salt, so much so that Russia became a salt-exporting country. Salt was to become the cornerstone of the Stroganov's fortune.

The records do not show how Luka rose above his fellow salt boilers. By the middle of the fifteenth century he had become rich enough to pay part of the ransom demanded for Moscow's Grand Prince Vasili the Blind by his Mongol captors. In those days, Moscow's ruler had not yet gathered all the Russian lands. Luka's loyalty stood to cost him dear if one of the Grand Prince's rivals won out in the struggle to rule Russia. He took the risk. He foresaw the day the Grand Prince would become the Tsar of All the Russias. In the centuries to come, the Stroganovs would serve as merchant adventurers, bankers, diplomats, courtiers and patrons of the arts. Always loyal, always ready to take risks in their sovereign's service, the Stroganov family stood as a pillar of the Russian throne till the Revolution of 1917. In return, the tsar showered these loyal subjects with honors and privileges making them the richest and most famous in Russia. Peter the Great made them counts of the Russian Empire. In the nineteenth century, a Stroganov married the daughter of the tsar.

In the 1520s and 1530s, Luka's grandson Anika, bought up the enterprises of his neighbors. He thus laid the foundation of that commercial empire which made the Stroganovs fabulously rich. He improved the salt-works. He scorned the narrow vision of Russian merchants and called in experts from Europe. These experts advised him to pump water saturated with salt from underground springs that flowed through subterranean salt beds and evaporate it in heated sluices. His vision paid handsome dividends. Anika's enterprises gave the Russians their first native source of salt large enough to meet their needs. The Stroganovs made a fortune from this salt monopoly.

But Anika was not any the more satisfied with the present even as he did not live in the past. He dreamed of empires yet to come. From salt he diversified into trading in furs. He traded his salt for the pelts of sable and black Arctic fox. He petitioned the tsar for the right to trade directly with the English merchants who found their way through the White Sea to the mouth of the Dvina River on Russia's northern coast. Anika offered the English furs which would be worth a fortune in the markets of Europe. In return, he received western luxuries. These luxuries he traded to his countrymen for more furs. Like men of ambition, he knew to curry favor with the powerful and strong. As the tsar's "humble slave", he sent the best of his furs to his royal master. This policy was to stand him in good stead in the long run. When Russia's conquest of Kazan opened the way to Siberia, he and his sons stood at the vanguard of those seeking to take advantage of the situation.

By that time, Anika was well-established in his base at Solvychegodsk. He built there a massive wooden castle. Measuring more than two hundred feet across its front, its heavy wooden walls were punctuated with tiny windows that let in only shafts of light. Torches and oil lamps cast shadows across its gloomy interiors even at midday. The castle was dark and dingy like the caverns that produced the salt that gave the wealth for its building. The Stroganov castle just fell short of being grotesque. Towers and turrets reminiscent of the style of Europe reflected Anika's efforts to draw closer to the West. This was at a time when a xenophobic church and government worked to preserve the old ways. At Moscow, travelers from the West were confined to a certain section of the city. Russians were taught to regard foreign ways and learning with deep suspicion. However, at Solvychegodsk, Europeans came and went frequently and at will since the liberal Anika saw things differently.

Anika saw foreigners as the key to build his empire. He insisted that his sons learn the ways of the English, the Dutch and Germans with whom his agents exchanged furs, forest products and grain. The west, he knew, had much to offer. They had the technology to sail the ocean at a time when the ignorance of the compass forced Russian mariners to hug the coast. Anika knew that whoever had the science of the West held the key to the future. He worked to bring the learning of the West to Solvychegodsk. He sought to integrate it in the world around him. His experiment in architecture proved awkward. The gap between the two cultures proved difficult to bridge. Nevertheless, the Stroganov castle held the largest private library anywhere in Russia. It became a melting pot for the culture and politics of the East and West. There, learning, wealth, security and success blended together to form a firmly set anchor that guaranteed his family's fortune. For another century, Solvychegodsk would prove to be the nerve center of the Stroganov's empire. The focus of this empire was shifting from Europe to Siberia, from West to East.

From the awkward base raised out of the north Russian forest, Anika presided over enterprises which stretched from the White Sea to Moscow and beyond. At fifty, there was a lot of the biblical patriarch about him. His ways of doing business and the way in which he ruled the lives of his three sons reminded one of an Old Testament Prophet. Of his three sons, Grigori inherited his love of learning. Yaakov was endowed this practical interest in science. Semen, the youngest had his father's thirst for beauty. In the sixteenth century's end, he was to sponsor and organize the famous Stroganov school of icon painters in Russia's Siberian borderlands. In the mid-1550s, Semen held the Stroganov bastion at Solvychegodsk while his father and brothers set out to conquer the Siberian frontier. Semen thus became the Stroganov patriarch for the next generation before his time.

The Russian conquest of Kazan opened new vistas for the Stroganovs. Their attention was drawn to the land called Perm. Even while the Tatars still ruled Kazan, Stroganov agents had explored the territory beyond it. They found new salt springs at Solikamsk. The land here was easier to till than that of the far north. In the Perm territory, whose rivers connected directly with Moscow, the Stroganovs could increase their salt production at lower cost. They could feed their workers cheaply. Moreover, they would have access to the fur-filled lands of Siberia. The Perm lands were a great prize to whoever could exploit them. The Stroganovs expected the Perm lands as reward for their long years of giving gifts and loyal service to the tsar. They had but to ask, and Perm was theirs. Ivan the Terrible hoped to kill two birds with one stone. Not only was Perm a reward for the loyalty of the Stroganovs, it was also an area which would be developed by the entrepreneurial skill of the Stroganovs. The wilderness could thus be transformed to the treasury's advantage. Allowing the Stroganovs to develop European Russia's eastern frontier was an investment in the future. Ivan's descendants would reap the dividends. The taxes that Anika's heirs would pay on their profits would add to the coffers of Ivan's successors in the years ahead. In the meantime, holding the eastern lands of the Perm territory gave the Stroganovs control of Russia's Siberian gateway. The Kama and Chusovaya rivers that flowed from it connected the crest of Ural Mountains directly to the Volga and the river route to Moscow.

In 1558, Ivan the Terrible leased nine and a quarter million acres of "empty lands, dense forests, wild streams and lakes, empty islands and pools" that lay "along both banks of the river Kama up to the river Chusovaya" to Anika's son Grigori. Grigori received full rights to develop the lands as he saw fit. He was to report to Moscow any gold, silver, copper or tin he found there. He was exempted from tax for the first twenty years. Grigori received a royal dispensation to build forts and "to set up in it canon and muskets, and station gunners, and musketeers and artillery men" to defend his salt works and settlements against native attacks from the Siberian side of the Urals. Here the Stroganovs reigned supreme, free even from the authority of the tsar's officials. "The tsar's officials cannot handle cases of the Stroganovs and their men" a supplementary charter stated in 1564, "unless the latter commit murder or are caught while plundering." Stroganov settlements and blockhouses thus became the Russian's staging area for the conquest of Siberia.

A decade after he had leased the lands along the Kama, Ivan the Terrible gave the same rights to the lands along the Chusovaya from its source in the Urals to the point where it emptied into the Kama to Yaakov. From there, the Stroganovs could send their trappers, traders and tribute collectors along the rivers of western Siberia. They could begin opening up the empire Anika had been dreaming of since his days as a salt-boiler in Solvychegodsk. The Stroganovs were now the masters of a region two-thirds the size of sixteenth century England. They had become the largest private landholders in Russia. Rewarded for their loyalty to a tyrannical tsar, they had risen in the space of one generation from humble salt merchants to Russia's richest entrepreneurs.

For Anika, the race for power and wealth was over. He was seventy when Grigori got his first charter from the tsar. The old man had outlived two wives. He was now ready to devote the last years of his life to prayer and contemplation. Leaving his sons and grandsons to carry on his work, the Stroganov patriarch took the name of Yasof and retired to a monastery. There he lived out his final days as a monk. His sons took the first steps to shift the eastern edge of the empire he had built into Siberia itself. Equal to their father in shrewdness and possessed of greater energy, Grigori and Yaakov Stroganov sent the first traders across the Urals to collect fur tribute from the natives and fill the Stroganov warehouses with rare Siberian pelts.

As agents of Yaakov and Grigori Stroganov moved the edge of the Stroganov empire into Siberia, their settlements along the Kama River bore the brunt of the first Siberian counterattacks. Almost directly east to the point where the Chusovaya River had its beginnings near the Ural's crest, the khanate of Sibir, the sister state of the fallen Tatar khanate of Kazan, stood guard over Siberia's western fur lands.

 The internecine feuds in the khanate of Sibir prompted Ivan the Terrible to demand thirty thousand sable pelts from them as yearly tribute. The Tatars sent seven hundred and despite continued threats from the tsar's diplomats never sent more than a thousand. Moscow, they knew, had become too deeply embroiled in a war with Sweden and Poland to enforce its demands for tribute in the east so long as they sent a token payment.

Kuchum Khan who seized power in 1563, brought to heel the rebel princes and feuding lords of the khanate. For the first time in half a century, the lands and lords of Sibir were under control.

Observing this debacle of the Russians in the Crimean war, Kuchum Khan stopped paying tribute. He hoped to take advantage of the native's resentment at the way in which the Stroganov's new settlers "seized the places where the natives hunted beavers and gathered honey and fished". He decided to raid the Perm lands of the Stroganovs.

By the time of Kuchum's first campaign, the Stroganov's frontier settlements held close to ten thousand freemen and another five thousand serfs. These men and women had built saltworks to boil the waters of Perm's salt springs. They had carved farms out of the wilderness. They had built blockhouses and stockades to protect their homes and fields. They were, in short, well-prepared for any attack that might come against them from the east. At first, Kuchum sent his native allies to raid the Stroganov settlements along the Kama. In July 1572, they killed nearly a hundred Russians. An enraged Tsar Ivan urged the Stroganovs to "live very much on your guard." He advised them to assemble as many men as they could muster to defend their lands against the Tatars. For natives willing to turn against the Tatars and join the Russians, he added, "We shall make all things easy for them."

The next year the Stroganovs faced a larger Tatar force under the command of Kuchum's nephew. This time not only were the Russians massacred but so were a number of natives who had joined them. Fearful that larger, better-equipped expeditions might soon follow, the Stroganovs petitioned the tsar for permission to carry the fighting into the heart of Kuchum's kingdom. Their petition reached Moscow not long after Ivan learned that one of his own envoys had been killed by Kuchum's forces. He extended the Stroganovs charters of authority into Siberia itself. He gave them permission to build frontier forts on the edge of Kuchum's lands. As the tsar's agents, the Stroganovs now had permission to invade Asia.

The Stroganovs decision to fight Russia's first war in Siberia was an act of great daring and audacity. Once in Kuchum's lands, the Ural

Mountains and several miles of trackless wilderness would separate their forces from their bases of support in Perm. This would make it difficult to send reinforcements and new supplies. What made the situation all the more uncertain was the tsar's second thoughts. The past quarter century had been war-torn. The Russo-Kazan and the Russo-Crimean wars had been exhausting. Ivan suffered from a crippling spinal disease. The stubbornness with which the Tatars had opposed him in Kazan and the Crimea made him wary of taking on Kuchum. He doubted if the Stroganovs commanded the manpower or the weapons needed to conquer Kuchum's kingdom. In the west, the war with Sweden and Russia dragged on and on. Ivan, in the late 1570s hesitated opening a second front in the east. Knowing that his commanders would call for more weapons and men to fight the Baltic wars, Ivan ordered the Stroganovs to pull back from Siberia.

Yet, the Stroganovs were not ones to fight shy of taking risks. After Grigori died in 1575 and Yaakov in 1579, their sons Nikita and Maksim continued the fight. They knew they could not hold the Perm lands if the attacks from Siberia became stronger and more frequent. Taking advantage of the great distances that separated Solvychegodsk and the Perm lands from Moscow, Nikita and Maksim Stroganov decided to ignore the tsar's orders and carry the war to Kuchum's capital. A victory in Siberia would boost the morale of tsar accustomed to defeat.

The decision of the Stroganovs was all the bolder, taking into consideration Ivan the Terrible's fits of rage which involved the ruin of favorites. If they lost, they lost more than a battle, they lost the tsar's favor. By the end of the 1570s, Anika's grandsons had started recruiting bands of Cossack to fight their war. Such adventurers had a long and inglorious history of banditry in the southern lands of Russia. Equally at home in land or water they had a reputation of being some of Russia's most daring fighters. If the Cossacks could work their way along the Kama and Chusovaya and portage across the Ural's crest, they could sail their high-sided boats into the heart of Kuchum's kingdom. In Yermak Timofeyevich, Maksim and Nikita, found a leader who matched in courage the daring of these men. A brigand with a checkered career, Yermak became the man who led the "conquest" of Siberia.

Chapter-6

The "Conquest" of the Khanate of Sibir

MYSTERY AND LEGEND SHROUD the image of Yermak Timofeyevich. Folklore and legend fortified with a modest sprinkling of fact make up the image that has come down to us. Like all swashbuckling conquistadors in this age of discovery and exploration, he combined brutality, cunning and daring. A frontiersman, he was a pioneer discovering new Promised Lands. A product of the age of Ivan the Terrible, Queen Elizabeth, and Sir Francis Drake, he combined the cruelty, craftiness and audacity of all three as he opened the way to a new world that none had ever dreamed of. Until he entered the service of the Stroganovs that spring, we only know that he fought for the tsar against the armies of Poland. As a brigand of renown, he plundered merchant ships upon which the tsar bestowed his protection. Beyond that not even one contemporary description of Yermak has survived. The Church Fathers in those days condemned the painting of portraits and we can only guess what he looked like. One Russian chronicle describes him as "flat-faced, black of beard with curly hair, of medium stature and thick-set and broad-shouldered." But that account was written more than a century after his death by a monk who had never seen him and did not even know the exact year in which Yermak's "conquest" had taken place.

Measured against the great hosts that conquerors from the east had led against the West in earlier times, Yermak's army of 840 men was a tiny force, but it was similar in size to those European expeditions that had carried out the great colonial conquests of the New World earlier that century. Only eighty-eight men had sailed with Christopher Columbus to discover the New World and Hernando Cortes had conquered the Aztec empire with a force of 617. Like the captains of Spain, Yermak would have the advantage of fighting with firearms against an enemy without them. The Ural Mountains barrier would cut him off from his bases of supply nearly as completely as the Atlantic Ocean did the expeditions that conquered the New World.

Nikita and Maksim Stroganov spent twenty thousand rubles to equip and supply Yermak's force. Against the bows, arrows and spears of Kuchum Khan's armies, Yermak's Cossacks carried matchlock muskets, sabers, pikes and several small canons. The war started through punitive skirmishes. In July 1580, 540 Cossacks under Yermak Timofeyevich invaded the territory of the Voguls, subjects to Kuchum, the Khan of Siberia. They were accompanied by 300 Lithuanian and German slave laborers, whom the Stroganovs had purchased from the tsar. Throughout 1581, this force traversed the territory known as Yugra and subdued Vogul and Ostyak towns. In May 1581, the campaign began in earnest. At this time, they also captured a tax collector of Kuchum. After a three-day battle on the banks of the river Irtysh, Yermak was victorious against a combined force of Kuchum Khan and six allied Tatar princes. On June 29, the Cossack forces were attacked by the Tatars but again repelled them.

Following a series of Tatar raids in retaliation against the Russian advance, Yermak's forces prepared for a campaign to take Isker the Siberian capital. Yermak set out for Siberia on September 1, 1581, from a base camp on the Chusovaya River, sailing up the river toward the Ural Mountains, passing beneath fierce cliffs and dense forests. Divided into companies of fifty, each with a supply of rye flour, cracked buckwheat, salt, gunpowder and lead, and with three priests and a runaway monk to give them spiritual solace, Yermak's force set out in a flotilla of high-sided boats.

After six days, his men dragged the boats ashore and began to carry them over a pass. Here, in the pass, they built a fortified camp where they remained throughout the winter, and in May they broke camp, lowered their boats on the Tagil River, and sailed eastward into the domain of Khan Kuchum, raiding Tatar settlements along the way. Once when they captured an important Tatar officer, they sent him ahead to Sibir where he reported to the Khan that the Cossacks had weapons that "spewed out flame and smoke with a sound like thunder." The muskets and the three cannon terrified the Siberians but they continued to fight.

Yermak fought a whole series of engagements on the way to Sibir. Progress was slow; many of his men were killed, and many more were wounded.

Throughout September 1582, the Khan gathered his forces for a defense of Qashliq. A horde of Siberian Tatars, Voguls and Ostyaks massed at Mount Chyuvash to defend against invading Cossacks. On October 1, a Cossack attempt to storm the Tatar fort at Mount Chyuvash was held off. On 23 October, the Cossacks attempted to storm the Tatar fort at Mount Chyuvash for a fourth time when the Tatars counterattacked. More than a hundred Cossacks were killed, but their gunfire forced a Tatar retreat and allowed the capture of two Tatar cannons. Then on October 26, 1582, the Cossacks fought the battle with Mehmet Kul, Kuchum Khan's nephew which was to change the course of Siberia's history. On the banks of the Irytysh, they met the Tatar charge with a volley of massed musket fire that wounded Mehmet Kul. Mehmet Kul had to be carried away to the farther bank of the Irtysh. This broke the spirit of his warriors. Shortly afterward, Kuchum Khan, who had been watching the battle from a hill some distance away, rode off to Sibir, not to put the town in a state of defense but to gather up his treasure before escaping eastward.

The Russians had lost about two hundred men in the previous engagements, and they lost over a hundred more in the battle for Sibir. Three days later, Yermak and his men occupied Isker to celebrate what has traditionally been called the "conquest of Siberia." They divided the wealth of sable and Arctic fox pelts that filled Kuchum's treasury. A few days later, the Ostyaks and Tatars began to return, having been promised that no punishment would fall on them. Yermak threatened his Cossacks with death if they so much as touched the local inhabitants. Yermak was well aware of the importance of his conquest, for the way was now open for the conquest of the rest of Siberia to the Pacific Ocean.

Then, burdened with their plunder, they settled down to spend their first winter in Siberia, with the river highways frozen over.

Though Yermak fought in the pay of the Stroganovs, he claimed Siberia in the name of Ivan the Terrible. He hoped to make amends for his earlier crimes as a river pirate. On December 22, 1582, Ivan Koltso, his second in command, set out from Sibir for Moscow together with fifty of the Cossacks. They took with them a larger number of furs and skins as presents for the Tsar-there were sixty bundles of sables, forty pelts to a bundle, twenty black fox skins and fifty beaver skins. They also carried long letters from Yermak addressed to Ivan and to the Stroganovs with a full account of their exploits. To the Tsar, Yermak wrote that he had undertaken the expedition to expiate his sins of brigandage, and now that he had succeeded in conquering the Khanate of Siberia, which will remain in Russian possession "for ever and ever, so long as God lets the world stand," he hoped and prayed for a free pardon for himself and his Cossacks. If no free pardon was granted, he promised to offer his life heroically on the scaffold.

The long letter which had taken two months to write, finally reached Ivan at the end of January or the beginning of February 1583. The church bells rang, and there were services of thanksgiving to celebrate the new conquest. In the streets of Moscow, the people could be heard congratulating one another with the words, "God has given a new kingdom to Russia." Everyone shared in the excitement, remembering that there had been little good news in Russia for many years. Ivan Koltso became the hero of the hour, feted by everyone, and welcomed in the throne room of the Kremlin

Palace, where he bowed before Ivan and kissed his hand. The misdeeds of the Cossacks were forgotten.

From the point of view of gaining the tsar's favor, Yermak's men reached Moscow at the best possible moment. The Russians had finally ended their war with Sweden and Poland. The tsar, freed from the Baltic wars, was ready to support the Stroganovs in the east. As a mark of favor, Ivan the Terrible sent Yermak a long coat of fine chain mail, emblazoned with the gilded double-headed eagles that formed the crest of Moscow's' rulers. He then added "Tsar of Siberia" to his many titles.

Siberia passed into possession of the Tsar. The Bishop of Volgda was commanded to send ten priests together with their families to Siberia. Ivan Koltso was authorized to seek new settlers for the conquered land. Prince Simeon Bolkhovsky was ordered to proceed at once with five hundred musketeers to Sibir to establish the Tsar's authority.

The Stroganovs too found themselves in high favor. Ivan summoned them to court, heaped praises on them, presented them with new lands and estates on the Volga and permitted them to trade without paying any taxes or duties in all their towns and villages. This was a princely gift, representing vast new wealth, a proper reward for having added a new empire to Ivan's vast territories.

About the same time that Ivan Koltso was being feted in Moscow, in February 1583, the Cossacks succeeded in capturing Mehmet Kul, the nephew of Khan Kuchum. Yermak received him kindly, but kept him under strong guard, for he was a brave and daring commander and had caused much havoc during the previous three months. When Ivan heard of the capture of Mehmet Kul, he ordered that the prince should be brought to his court, received him well, and according to the long-established custom granted him the rights enjoyed by all Tatar princes who swore an oath of loyalty to him and entered his service.

Meanwhile Khan Kuchum roamed the steppes of central Asia,

determined at all costs to regain his lost kingdom, continually plotting. From his base upriver, he sent raiding parties to harass the Cossacks. Although Yermak continued to win the larger battles against the remnants of Kuchum's armies, ambushed by the Tatars and their Ostyak allies continued to whittle away his tiny force. To live through a second winter in Siberia, he needed the supplies, gunpowder and shot that the tsar had sent through his musketeers. These failed to arrive before the snow fell, leaving Yermak with only the handful of men who had crossed the Urals with him a year before. When the summer of 1584 brought no reinforcements the Cossacks knew their chances of surviving a third Siberian winter were slim.

The tsar's musketeers finally reached Yermak's camp in November 1584, more than a year late. They had used up or lost all the supplies brought with them. The Cossacks had failed to stock up enough provisions for the winter. The result was starvation. The amount needed to feed so many was not forthcoming. Their situation was all the more desperate once the Siberian winter struck. "Many people died", one of the chronicles reported. "Men were forced to eat the bodies of their companions who had died from hunger."

Hunger and scurvy took their toll. Tatar ambushes claimed a large share of the survivors. Once conquered, the Tatars often rebelled. They had many advantages: they outnumbered the Russians, they traveled lightly, and they did not suffer from scurvy, which killed off half of Yermak's Cossacks. An uprising led by a certain Karacha spread across Siberia, and Sibir, now a heavily defended fortress town, was surrounded by Karacha's forces. On the night of June 12, 1585, the Cossacks crept out of the town, attacked Karacha's forces while they were sleeping, killed many of them and sent the rest fleeing across the Irtysh River. The Tatars were quite capable of using the same trick. On August 5, less than two months later, Yermak himself was resting in his camp only two day's march from Sibir when he was surprised by a force commanded by Khan Kuchum. During the night it rained heavily, a strong wind rose, and Yermak gave his men permission to sleep, for it was inconceivable that the Khan should attack on such a night. He did not know that the Khan's spies had been following him closely and knew exactly where he was. In the middle of the night, Khan Kuchum's Tatars fell on the camp, which was set in ancient Tatar burial ground and slaughtered all the Cossacks except for two men who succeeded in escaping. One of them succeeded in regaining Sibir. The other was Yermak, who dived into the swift-flowing Irtysh River in full armor, hoping to reach the place where the Cossack boats were moored. But weighed down by the armor emblazoned in gold with the double-headed eagle, he drowned. On that same night, forty-eight of his Cossacks were killed in the burial ground.

Thus died Yermak, the thickset, broad-shouldered, black-bearded conqueror of Siberia, having ruled over his conquest for less than three years. Dead, he entered into legend. Songs were sung and stories were told about him. Some of them had little enough relation to historical truth but were all the more truthful for being imaginative. After the conquest of Siberia, he never returned to Russia, but the popular imagination demanded that he should be seen confronting the terrible Tsar, and so they sang:

Then Yermak and all the Cossacks departed from Sibir
And made their way to the stone city of Moscow
To the terrible Tsar Ivan Vasililievich,
And the Lord Tsar spoke to him, saying:
"Well, Yermak Timofeyevich, where have you been?
How many people have you been robbing?
How many innocent souls have you killed?
How did you capture the Tatar Khan?
How did you bring the Tatar army into my power?"

Then Yermak fell to his knees,
Bowing low before the terrible Tsar,
And thrust a letter into the Tsar's hands
And with the letter went the accompanying words:
"before you, O Lord, I profess my guilt.
We have been brigands on the blue waters!"
And the Lord Tsar was not in the least angered,
And showed great mercy to the sinner,
And ordered him to be loaded with gifts.

Then Yermak was sent back to the land of Sibir
With orders to collect tribute from the thieving Tatars
So that the Tsar's treasury should be filled up,
And he returned with his Cossacks to Siberia
To fulfill the orders of the Tsar
To extract from the thieving Tatars
The necessary tribute and to bring them
Even more under the Tsar's authority,
And they did so with the utmost eagerness.

For the Russians, Yermak became a folk hero, the conqueror not merely of a small Tatar kingdom but the whole of Siberia. Over the years, the historical folk songs from which the Russian masses drew comfort transformed this one-time river pirate into a gallant knight, a champion of the unfortunate and a hero who knew no reproach. Yermak became Russia's King Arthur with touches of Roland, Siegfried and Sir Galahad. Among the Tatars, Yermak became a brave spirit who could heal the sick and ensure victory to the downtrodden. Tatar legends told how his body had remained uncontaminated for many days after Kuchum's men had pulled it from the river and how, after they secretly buried their fallen foe on the riverbank, columns of fire visible only to Siberians marked his grave by night. One Siberian ally begged the tsar for Yermak's coat of mail believing it would make him invincible.

In 1586 the Russians returned, and after subduing the Khanty and Mansi people through the use of their artillery they established a fortress at Tyumen close to the ruins of Qashliq. The Tatar tribes that were submissive to Kuchum Khan suffered from several attacks by the Russians between 1584–1595; however, Kuchum Khan would not be caught. Finally, in August 1598 Kuchum Khan was defeated at the Battle of Urmin near the river Ob. In the course of the fight the Siberian royal family were captured by the Russians. However, Kuchum Khan escaped yet again. The Russians took the family members of Kuchum Khan to Moscow and there they remained as hostages. The descendants of the khan's family became known as the Princes Sibirsky and the family is known to have survived until at least the late 19th century.

Despite his personal escape, the capture of his family ended the political and military activities of Kuchum Khan and he retreated to the territories of the Nogay Horde in southern Siberia. He had been in contact with the tsar and had requested that a small region on the banks of the Irtysh River would be granted as his dominion. This was rejected by the tsar who proposed to Kuchum Khan that he come to Moscow and "comfort himself" in the service of the tsar. However, the old khan did not want to suffer from such contempt and preferred staying in his own lands to "comforting himself" in Moscow. Kuchum Khan then went to Bokhara and as an old man became blind, dying in exile with distant relatives sometime around 1605.

The gateway to Siberia Yermak had opened never closed. Even before news of Yermak's death reached Moscow, Boris Godunov, the regent for Ivan the Terrible's sickly son Tsar Fedor, had sent another force to help stake out the Russian's claim by building small frontier forts along the river routes that Yermak followed into Siberia. Almost every year after that more expeditions added new sites to the list of fortified places that were beginning to establish the Russian's claim to Siberia. In order to subjugate the natives and collect yasak (fur tribute), a series of winter outposts (*zimovie*) and forts (ostrogs) were built at the confluences of major rivers and streams and important portages. Obskii Gorodsk was built in 1585. Tyumen and Tobolsk followed — the former built in 1586 by Vasili Sukin and Ivan Miasnoi, and the latter the following year by Danilo Chulkov. Tobolsk would become the nerve center of the conquest. To the north Beryozovo (1593) and Mangazeya (1600-01) were built to bring the Nenets under tribute, while to the east Surgut (1594) and Tara (1594) were established to protect Tobolsk and subdue the ruler of the Narym Ostyaks. Of these, Mangazeya was the most prominent, becoming a base for further exploration eastward. By 1600, Moscow's musketeers and Cossacks had claimed all of the territory between the Urals and the Ob. They were beginning to reach toward the Yenisei, the next of Siberia's great rivers which lay nearly a thousand miles to the east.

Ahead of the Russians stretched a vast domain, the breadth and limits of which no one knew or imagined. Fewer than two hundred thousand natives lay scattered across Siberia's five and a third million square miles. Some were hunters and gatherers while others lived as nomads. Few tilled the soil. Siberians of the north had come to grip with the emptiness of the tundra its barrenness broken by only a few dwarf bushes. In the forests of the taiga, the natives lived by hunting and fishing. The Siberians of the south lived as nomads, herding cattle and horses across Eurasia's million square miles of grasslands.

Without firearms these Siberians could not oppose Russia's Cossacks. As they moved ahead, fighting climate, terrain and distance these Russians made the Siberians into payers of tribute, demanding from each a rich yearly ransom in furs and wreaking terrible vengeance on any tribe or clan that refused. The Russian conquest of Siberia was accompanied by massacres due to indigenous resistance to colonization by the Russian Cossacks, who savagely crushed the natives. When the Russians did not obtain the demanded amount of yasak from the natives, the governor of Yakutsk, Piotr Golovin, who was a Cossack, used meat hooks to hang the native men. In the Lena basin, 70% of the Yakut population died within 40 years, and rape and enslavement were used against native women and children in order to force the natives to pay the Yasak. Siberians had faced such oppressions before, for some of Asia's most warlike tribes had imposed similar burdens upon their weaker neighbors in the year's after the Mongol's decline. Yet, while their former oppressors had tempered their demands with the understanding that ruined people could not pay tribute, the Siberian's new masters measured the value of their subject's lives only in terms of the furs they could squeeze from them.

Before the Russians came, Siberia's natives had lived in harmony with the world around them. They never exhausted the supply of wildlife upon which their survival depended. They moved their flocks at proper intervals so as not to overgraze the steppe. They suffered natural disasters but never created man-made ones. The coming of the Russians replaced that harmony with a system of exploitation based upon greed and arrogance. Draining one Siberian resource after another, the Russians, over the next four hundred years would turn Siberia into an ecological catastrophe.

Arrogantly, with all the self-righteous certainty of people convinced that they had come to confer the blessings of civilization on other peoples of the world, the Russians dated the beginnings of Siberia's history from October 26, 1582, the day of Yermak's conquest. Yet Yermak had not entered a world without a past of its own. For the 175,000 men and women whom the forces of history scattered across Siberia, "the beginnings" stretched back to a dimly remembered time. This age was enshrined in the deeds of folk heroes and great founders who had brought new peoples into being. For most of Siberia's past there is no written record. Only legends, folklore and remnants of people and places remain to tell the story of Siberia's past.

Chapter-7

The Siberians

HUMAN INHABITATION IN SIBERIA dates from prehistory. The first settlements in Siberia began around 10,000 B.C. Prehistoric Siberians organized their lives around the mammoth which served as their chief source of food. They built their homes from mammoth femurs many feet below the ground to shield them from Arctic winds. They shaped flint like the cave dwellers of Paleolithic Europe and like the Eskimos of North America. They called upon shamans to communicate with the dead and with the great forces of nature. In these ways, they came to terms with one of the most brutal climates anywhere on earth.

Then the climate of the earth warmed. The great Eurasian glaciers retreated. The mammoths died out. Siberia's inhabitants had to reshape their lives around herds of cattle, horses and reindeer. The advent of agriculture and animal husbandry accelerated the pace of progress. The Afanasevo culture emerged in south Siberia, occupying the Minusinsk Basin and the Altai Mountains in 3500-2500 BC. The Afanasevo people discovered how to produce bronze from the copper they extracted from the dull green rocks that Nature had scattered across the hills around them. During the next millennium, the Afanasevo people evolved into the more sophisticated Andronovo societies in present-day Kazakhstan and the eastern Urals to the region west of Lake Baikal.

Though ahead of their neighbors in the taiga and tundra, the Andronovo remained behind the civilizations that flourished in Egypt, Babylon, China and India. While the Egyptians built the pyramids and Hammurabi framed his code of laws and the Bronze Age flourished in Shang China, nothing even remotely similar emerged in Siberia. However, its southern parts were drawn into interaction with the East and West.

Life in the ancient world moved at snail's pace. However, by the first millennium B.C., trade was beginning to bind its outlying parts. Sea routes connected the ports of south China with India, Persia and Egypt. Overland routes through Sinai connected Egypt with the rest of the ancient world. Other overland lanes connected the Euphrates valley with the Mediterranean ports of Antioch and Alexandria. Further north, the great Silk Road stretched across Eurasia from Luoyang in north China through the oases of Turkestan and the empire of the Parthians to Baghdad and Antioch. Along this route moved slow moving pack horses and two-humped Bactrian camels. They carried sandalwood, aromatic resins, silks and spices from the East to the West and exchanged them for the luxuries of Persia, Greece and Rome. The Andronovo people of south Siberia played a part in this trade. They sent their furs to the bazaars of Central Asia where they traded them for the treasures and luxuries produced by the sophisticated civilizations East and West. This trade provided the rich pelts that lined the silken coats of Chinese mandarins and Russian aristocrats. Their furs commanded a king's ransom in the cities of China, Bactria, Persia and Rome. Thus, the people of Siberia were closely tied to the world's great centers in the days of Alexander the Great and Julius Caesar.

The burial mounds of the Pazyryk people testify to the prosperity of south Siberia those days. This wealth is evident in the wide array of finds from the Pazyryk tombs, which include many rare examples of organic objects such as felt hangings, Chinese silk, the earliest known pile carpet, horses decked out in elaborate trappings, and wooden furniture and other household goods. These finds were preserved when water seeped into the tombs in antiquity and froze, encasing the burial goods in ice, which remained frozen in the permafrost until the time of their excavation. The most famous undisturbed Pazyryk burial so far recovered is the Ice Maiden or "Altai Lady" found by archaeologist Natalia Polosmak in 1993 at Ukok, near the Chinese border. The find was a rare example of a single woman given a full ceremonial burial in a wooden chamber tomb in the fifth century BC, accompanied by six horses. Near her coffin was a vessel made of yak horn, and dishes containing gifts of coriander seeds: all of which suggest that the Pazyryk trade routes stretched across vast areas of Iran. Golden jewelry and ornaments in the shape of stylized deer, tigers, horses and winged griffins show that the art and culture of East and West had made a mark upon life in Siberia's south-central highlands in the centuries before the advent of Christ. This was a world different from the isolated one the Russians found two thousand years later.

The stirring of nomadic peoples destroyed the harmony established by the ancient Silk Road. These nomadic peoples inhabited the steppes along Siberia's southern frontier. They were to destroy the prosperity of Siberia and push it back into an age of darkness. Around 700 B.C., the Scythians burst out of their homeland on the edge of present day Mongolia. They began the great westward campaign that was to plunge Eurasia's great grass road into two thousand years of turmoil. After the Scythians came the Sarmatians and then the Huns followed by the Mongols. Each of their massive migrations weakened the ties that bound the Altai and Minusinsk lands with the East and the West. Once on the verge of being drawn into the mainstream of world trade and civilization, Siberia began to drift away. It turned inward into itself and fell rapidly behind. Siberia's fur trade shifted, first to China, and then when the Christian era got under way, north to Russia.

In the dark centuries when the Mongols overran Eurasia, Siberia was plunged into darkness and obscurity. The only exception were the Kirghiz people. The Kirghiz were originally masters of the lands between the upper reaches of the Yenisei River and Lake Baikal. They overthrew the Uighur khanate in A.D. 840 and ruled over Mongolia in their stead for about three-quarters of a century. Driven back to their former lands around A.D. 920, they settled down in the corner of the steppe that separates the Tien Shan from the eastern shores of the Caspian Sea. Those who remained around the upper

Yenisei submitted voluntarily to Chinghiz Khan in 1207.

Aside from the Uighurs, the Kirghiz were the only Siberians who could read and write. They tended not only flocks but fields, turned the ground with iron-tipped plows, and raised millet, barley and wheat. They employed slave labor to build stone-paved roads. They built extensive irrigation canals to make the deserts of Central Asia bloom. By the middle of the ninth century, they were trading furs to the Arabs for cloth. However, they saved their finest pelts for trading with the Chinese for luxury goods. Kirghiz smiths crafted fine ornaments of gold and silver. They forged excellent iron weaponry. These had a market in China. Kirghiz blades were highly valued.

The Kirghiz had voluntarily submitted to Chinghiz Khan. Their compliance was rewarded by being allowed to continue their mining and iron works and their continuance of tending their flocks. With the help of Chinese craftsmen, they also wove brocade and crepe from silk. But as the Mongol empire broke apart so did the Kirghiz way of life. Within the space of two hundred years the Kirghiz people lost their written language. They gave up their trades and abandoned agriculture for their earlier nomadic way of life. By the seventeenth century, they became deeply embroiled in tribal and clan wars. Some of them greeted the Russians as a way of preventing political and economic chaos from overwhelming them. One master was better than several and one tribute collector better than many.

Although Mongol domination weakened the Kirghiz, it left the Buryat strong. The Buryats were distant relations of the Mongols. They ruled the forests and upland pastures to the Mongol's north. Their way of life was akin to that of the Mongols. The Mongols never conquered the Buryats but drove them further north. They continued to tend their herds and flocks in the rough lands of the Transbaikal. For many centuries, shamans ruled the lives of the Buryats. They required a horse to be sacrificed at each warrior's burial. Eventually, Buryat life changed affected by influences from the south. In the 1640, the Buryats were replacing shamans with lamas. Nevertheless, shamanism remained strong.

The Buryats never missed an opportunity to strike against the Russians. Buryat uprisings flared up around Lake Baikal several times during the seventeenth century. Buryat arrows took the lives of tsarist tribute collectors. Only in the middle of the eighteenth century did Buryat and Russian intermingle with Russian men marrying Buryat women.

The Tungus who lived in Manchuria were closely related to the Manchu lords who ruled China. Political upheavals in China uprooted the Tungus with the warriors of Chinghiz driving them farther north. There, in the taiga, the Tungus became hunters and reindeer herders. They moved with the seasons. The reindeer provided them food and transportation. They supplemented the meat and milk of the reindeer with the flesh of animals they hunted. Like other Siberian natives, the Tungus too relied upon shamans to communicate with their gods and to guide their lives. They interred their dead above ground by sewing them in reindeer skins. Then they set the hide-wrapped corpses atop high posts because they dared not violate the domain of the spirits that lurked beneath the forest floor.

"Cheerful under the most depressing circumstances, persevering, open-hearted, trustworthy, modest, yet self-reliant," the Tungus proved to be as easy for the Russians to conquer as the Buryats had been difficult. The westernmost Tungus paid their first fur tribute to the Russians in 1614. Within forty years, their eastern cousins were doing the same. Yet these easily conquered wandering natives proved resistant to modern ways. Right up to the twentieth century, all but a handful of the Tungus remained herders and hunters. They resisted settlement, farming and modernization.

Some of the Tungus adopted the ways of the Yakuts, Turkic-speaking Siberians settled along the Lena Valley. The Yakuts were iron-makers like the Buryat. They built their lives around raising cattle and horses. They supplemented the meat and milk their herds supplied with the game, fish and roots gathered from the taiga and tundra. Since they did not know how to make cloth, the Yakuts cured hides and skins crudely, smearing them with oil and smoking them in earthen stoves. Blacksmithing the ore they dug from river marshes was their only trade. They made some crude pottery shaped without a wheel.

The Yakuts were more numerous than the Tungus and more fecund. They had once been very powerful. They were weakened by tribal conflicts and clan rivalries. This made them easy to conquer. The Russians had to put down four bloody Yakut rebellions in the 1630s before they had these proud people firmly under control. Then the Yakuts, who had been shamanists like the Tungus converted quickly to Christianity. Ready to adopt Russian ways, they continued to live in yurts made of hides or birch bark in summer and a mixture of clay, bark and cattle dung in winter. Yakut yurts became notorious for their stench since they were closely connected to the sheds which protected their cattle from the Arctic winter.

West of the Yakuts and the Tungus, the Ostyaks herded reindeer, hunted, trapped and fished along the Ob and Irtysh rivers. The Ostyaks were related to the Finns and Hungarians by language. They lived in large families but did not organize into tribes as did their eastern neighbors. Those among them who lived in the tundra followed their reindeer herds. However, most of the Ostyak lived on the bounty of the forest. They hunted moose, reindeer, bear and fowl for food. They trapped sable, fox, squirrels and rabbits for their pelts. Of the animals they killed, hey ate the kidneys, liver and marrow raw, drank the blood fresh and dried or boiled the meat. In summer, they set aside hunting and trapping for fishing, using complicated traps to take large catches that they dried or ate uncooked.

Ostyaks engaged in trade with trappers and traders from Novgorod two hundred years before the Mongols came. They traded furs for iron tools and weapons. When Moscow replaced Novgorod in Russia's far north, its agents kept the trade with the Ostyaks going. Ostyak furs made their way to the Stroganov warehouses. When Yermak's conquest allowed the tsar to claim Siberia, the Ostyaks were the first from whom his Cossacks claimed fur tribute. At first, the Ostyaks paid. Then, when the tsar's officials demanded more furs as "gifts" in addition to the ten sable pelts per head required in tribute every year, the Ostyaks rose up in revolt. Ten years after

Yermak's victory, they besieged Berezov. They were brutally suppressed. They then attacked their Vogul neighbors. They attacked Berezov again in 1607 and the next year, tried to unite all of Siberia's natives against the Russians. They resisted Russian control for the better part of thirty years.

The Russians faced similar opposition fifty years and five thousand miles later. In 1640, the Koryaks of Siberia's northeast allied with their warlike Chukchi neighbors against the Russians. For more than a hundred years, the Koryaks fought the Russians. They recognized Russian sovereignty only in 1712. They continued to fight the tsar's tribute collectors for a half century after that. Even deeper in Siberia's northeast corner, on the Chukhotka peninsula, the Chukchi never did submit. The Russians built a strong fort at Anadyr at a cost of a million rubles. They collected barely a fiftieth of that cost in furs. In 1764, they therefore decided to leave the Chukchi to themselves. A hundred and twenty-five years later, the Chukchi finally agreed to pay a token tribute of 247 rubles to the Russian government every year.

The Chukchi and the Koryak were closely related by language and custom. They built their lives around hunting and fishing. The economy of the Chukchi and the Koryak combined elements of reindeer herders farther west with that of Eskimos on the other side of the Bering Strait. In earlier times, kayaks made of seal and walrus skin seem to have carried native Americans and early Siberians back and forth across these narrow waters. Chukchi and Koryak tools, weapons and ways of life bore striking similarities to those of Tlingit Indians in Alaska. In both cultures the skin of certain whales eaten raw with its layer of pink blubber still attached was thought to be a great delicacy.

Whether dealing with tundra reindeer herders, or steppe nomads, or its taiga hunters, the Russians thought only of furs as they crossed the five thousand miles that lay between the Urals and the Bering Strait. The astronomical prices that Siberian furs fetched in the markets of Moscow made rich men out of the first Siberian explorers. Others were left to tell of the fabulous wealth that lay in the lands beyond the Urals. Working in groups as small as two or three and as large as sixty, these men-called *promyshlenniki* in Russian-plunged into the Eurasian wilderness for years at a time before they returned laden with "soft gold." Kept alive by a raw courage that few possessed, the conquerors of Siberia sought the pelts of squirrels, foxes, ermines, martens and above all sable.

Described by an expert as "an arboreal marten native to the taiga of Eurasia......with fur ranging from brown to black, the sable, *Martes zibellina*, was about twenty inches long, lived alone and fed on everything from squirrels and grouse to pine nuts, berries and insects. A dense silky undercoat protected it against Siberia's ferocious winters. This formed its chief attraction for the wealthy men and women from Europe to China who clamored for its fur. "A beast full marvelous and prolific, "one seventeenth century Russian diplomat wrote, the sable was "a beast.... that the ancient Greeks and Romans called the Golden Fleece." In Siberia, this small animal, the size of a house cat was the magnet that pulled the Russians across the entire Eurasian continent by 1650.

Chapter-8

Spanning the Breadth of Asia

THE CENTURY AFTER YERMAK" S DEATH saw the rapid advance of the Russians across the Eurasian continent. The men who undertook this adventure were driven by a love of danger, a passion to explore the unknown and raw, festering greed. "they behaved as if this land were in imminent danger of falling to an enemy forever, and thus they had to take out everything that the enemy might be able to use," wrote Valentin Rasputin, "They found that vast Siberia, larger than any continent, became much less large and bottomless if they did nothing but scoop everything out." Literally, they scooped everything out. Sables, martens and foxes they took in the hundreds of thousands, squirrels they collected in the tens of millions. Almost every decade during the first half of the seventeenth century, the Russian's greed for furs pushed Siberia's fur frontier a thousand miles to the east. In order to subjugate the natives and collect yasak (fur tribute), a series of winter outposts (*zimovie*) and forts (ostrogs) were built at the confluences of major rivers and streams and important portages. The first of the outposts were Tyumen and Tobolsk — the former built in 1586 by Vasili Sukin and Ivan Miasnoi, and the latter the following year by Danilo Chulkov.

Tobolsk would become the nerve center of the conquest. To the north Beryozovo (1593) and Mangazeya (1600-01) were built to bring the Nenets under tribute, while to the east Surgut (1594) and Tara (1594) were established to protect Tobolsk and subdue the ruler of the Narym Ostyaks. Of these, Mangazeya was the most prominent, becoming a base for further exploration eastward.

River Routes: Three large rivers flowing from south to north, their tributaries stretching east and west like the intersecting branches of three mammoth trees, divide Siberia into large segments that the Russians crossed in several large bounds, leaping from one river basin to the next. Advancing up the Ob and its tributaries, the ostrogs of Ketsk (1602) and Tomsk (1604) were built. Ketsk *sluzhilye liudi* ("servicemen") reached the Yenisei in 1605, descending it to the Sym; two years later Mangazeyan *promyshlenniki* and traders descended the Turukhan to its confluence with the Yenisei, where they established the *zimovie* Turukhansk. By 1610 men from Turukhansk had reached the mouth of the Yenisei and ascended it as far as the Sym, where they met rival tribute collectors from Ketsk. To ensure subjugation of the natives, the ostrogs of Yeniseisk (1619) and Krasnoyarsk (1628) were established. The conquest of the Yenisei was complete by 1620. By 1630, they had reached the Lena.

Following the khan's death and the dissolution of any organized Siberian resistance, the Russians advanced first towards Lake Baikal and then the Sea of Okhotsk and the Amur River. However, when they first reached the Chinese border they encountered people that were equipped with artillery pieces and here they halted.

The Russians reached the Pacific Ocean in 1639. After the conquest of the Siberian Khanate (1598) the whole of northern Asia - an area much larger than the old khanate - became known as Siberia. Thus by 1640 the eastern borders of Russia had expanded more than several million square kilometers. In a sense, the khanate lived on in the subsidiary title "Tsar of Siberia" which became part of the full imperial style of the Russian Autocrats. Twelve hundred miles and nine years later they were on the shores of the Bering Strait, less than a hundred miles from Alaska's Cape Prince of Wales. Now at the easternmost tip of Eurasia, the Russians stood closer to the mainland of North America than had Columbus when his tiny flotilla first sighted land in the Caribbean.

Cossacks, government officials, traders and *promyshlenniki* played a part in the Russian conquest of Siberia. The conquest, in fact, was very complex with its various elements interdependent. At different times and under quickly changing circumstances, officials, traders and *promyshlenniki*, all played prominent parts. On some occasions, they responded to the vision of the tsar's chief commanders and governors in Siberia. At other times, they followed orders sent from Moscow, and in some cases they searched for fame and fortune on their own. Without significant logistical support from one of the tsar's military governors, trappers and traders found it next to impossible to amass the resources needed to explore Siberia's remote parts. Of necessity, traders, *Promyshlenniki,* and government men therefore squabbled but worked together. At times, there was more conflict among various military governors, each of whom compared for recognition in Moscow, than there was between them and the leaders of privately planned expeditions. On other occasions, governors who saw every independent expedition as a challenge to their supreme authority worked against the men who struck out on their own. One thing remained certain: Moscow's grip upon Siberia's conquerors weakened as the frontier moved east. It took the better part of a year for a messenger to travel from the tsar's palace in Moscow to the governor of Yakutsk and even longer to reach Okhotsk.

Experts agree that government leadership and soldiers sent from Moscow played the leading part before the Romanovs rose to power in 1613. This support and leadership from Moscow came at a time when the Russian state was engulfed by political turmoil following the death of Tsar Fedor in 1598. Several pretenders gathered armies along old Russia's frontiers. At one time, Poland occupied Moscow for almost a year. Yet, in this age of muddled European politics, the conquest of Siberia continued. Tsars, magnates and men who make pretenders into sovereigns- all sensed that Siberia's wealth would shape whatever future would unfold once the Time of Troubles was over. Siberia's wealth came up to a tenth of the royal Treasury's receipts during the seventeenth century. This would help the Romanovs rebuild Russia.

Built on the eastern side of the Urals, Verkhoture became the Russian's gateway to Siberia until the first segments of the Great Post Road from Moscow to Irkutsk was opened in the 1760s. Founded in 1598, Verkhoture held the tsar's customs house. All goods and furs going across the Urals had to pass this way for nothing passed duty-free before the 1770s. All winter-long, trains of sledges worked their way across the Urals, bringing flour, gunpowder, lead and salt from European Russia to be stockpiled at Verkhoture until the spring thaw opened Siberia's rivers. By the early 1600s, ninety river barges, each carrying up to thirty-six tons of supplies being sent on to Russian outposts on the lower reaches of the Ob and Yenisei rivers, left Verkhoture every spring.

While Verkhoture served as the gateway through which furs, supplies and trading goods passed between Russia and Siberia, Tobolsk became the nerve center of the conquest. After it became the headquarters of Siberia's first military governor in 1588 and the home of its first archbishop in 1621, power and wealth flowed to and from Tobolsk in all directions. By the 1650s, its merchants were selling over eighty thousand ruble's worth of goods a year and transshipping tens of thousands more. No other town east of the Urals played so large a part in Russian and Siberian affairs in the seventeenth century. One out of every four people who lived there worked in trade, one out of every two served in the army or government.

Russia's conquest of the lands between the Ob and the Yenisei proceeded along two roughly parallel east-west courses that lay roughly five hundred miles apart. The northern route went through the tundra, marking its centers at Berezov (on the Ob, founded in 1593), Mangazeya(founded above the Arctic Circle on the Taz River in 1601), and Turukhansk(founded in 1607 just below the Arctic Circle on the Yenisei).Using these centers as their headquarters, Russian Cossacks, working in squadrons of ten to thirty, ranged back and forth across the tundra, collecting fur tribute from the Ostyak hunters and Samoyed reindeer herders. At the same time, traders and *promyshlenniki* brought foreign goods from warehouses that British and Dutch agents had built at the newly opened White Sea port of Arkhangelsk to trade with the natives for more furs. Yet the end purpose of their efforts was always gold. A nation without significant gold and silver mines, old Russia needed to exchange Siberia's "golden fleece" for the hard metal that all nations had to acquire to be strong. Without Siberia's furs to convert into gold, Russia's poverty in the seventeenth century would have been astounding.

For something more than a decade beginning in the early 1600s, daring Russian seamen sailed through the Arctic seas from Arkhangelsk to Mangazeya, always hugging the coast, never letting land out of sight, but taking only a fifth as much time to make the voyage as was needed to travel overland and by river through Verkhoture, Tobolsk and Berezov. But if Russians could sail from Arkhangelsk to Mangazeya without compasses in four to five weeks, foreign captains with modern instruments and better ships might do so even more easily. There are documents that show that James I of England was planning to seize the tundra lands around Mangazeya around 1613. Fearful that foreign traders would use the Arkhangelsk-Mangazeya sea route to build their own direct links with the Siberian fur trade, the Romanovs closed it in 1619. After that the eclipse of Mangazeya as the tundra capital became only a matter of time for the overland route to it was too hard and dangerous to make it profitable.

Cut off from the outside world for nearly six months out of every twelve, Siberia's Arctic frontier towns flourished during the first half of the seventeenth century and then began to wither as the Russians learned that the barren tundra, without wood for fuel or shelter, could not support their eastward march. By the 1720s, Berezov had become very remote. Turukhansk became a place where political exiles were sent. Mangazeya, famous for its fair to exchange furs, had all but disappeared by the end of the seventeenth century. Except for a few charred logs left over from a fire in 1660s, nothing of it remained.

While Cossacks, promyshlenniki and traders struggled to keep Russia's tundra outposts during the first two decades of the seventeenth century, others advanced directly east from Tobolsk. They worked their way directly east along the tributaries that flowed into the Ob. After reaching Surgut in 1594 and Narym in 1598, Cossacks and soldiers following this southern route founded Tomsk on the Tom in 1604 and Makovsk on the Ket in 1618. Sometimes as at Tomsk they built a fort to help a native prince defend his lands against his Kirghiz enemies. The Russians took part in native wars sometimes. They fought against Ostyak, Tungus, Tatar and Kirghiz tribesmen. They relied on their small cannon and muskets to terrify their foes into submission. Native rebellions continued to plague the Russians. By the time they founded Yeniseisk in 1619, they held all the lands between the Urals and the Yenisei. Yeniseisk marked the point where the northern and southern routes of conquest intersected. After that they moved apart once more.

Yeniseisk was founded a thousand miles east of Tobolsk where the Angara flows into the Yenisei. It was to become the staging area for expeditions that would conquer Siberia's eastern lands. The tsar sent a military governor to rule Yeniseisk just four years after the Cossacks had raised the first logs of its palisades. Until the 1660s, it stood as the main bastion of Russian authority east of Tobolsk. From Yeniseisk, the Russians built a series of frontier forts to the southeast. In 1631 came Bratsk, followed by Verkholensk in 1641 and Barguzin in 1648. Barguzin was the point of departure for Russia's conquest of Transbaikal. The Cossack captain Kurbat Ivanov led an expedition of twenty-five promyshlenniki to stake Russia's claim to Eurasia's largest lake.

Despite their successes in the southeast, the main thrust of the Russian conquest lay directly east toward the Lena, the third of Siberia's massive rivers. Leading a detachment of ten Cossacks Vassili Bugor reached the Lena in 1628. He became the first Russian to collect tribute from the Yakuts living along it. Bugor returned to Yeniseisk without building a base for further exploration upon the Lena. Piotr Beketov, a Cossack lieutenant thereupon set out by a different route to bring to heel the new lands discovered. He writes to the Tsar Mikhail Fedorovich, "I, your slave…sailed down the great river Lena and, having arrived at the Yakut lands, built a small fort, and made all necessary defenses for it." He adds, "I shed my blood for you, Sire and defiled my spirit and ate mare's meat and roots and fir bark and all kinds of filth." Thus, in the flat official language of the time, Beketov reported the founding of Yakutsk by his force of thirty Cossacks in 1632.

Yakutsk was set on the right bank of the Lena halfway between its source and its mouth. It was an incredibly rich source of furs for the Russians. The natives in the territories round it placed a low value upon the sable pelts they supplied. They were willing to pay for a copper kettle supplied by the Russians by filling it with sable pelts. A decade after the setting up of Yakutsk 15The natives in the territories round it placed a low value upon the sable pelts they supplied. They were willing to pay for a copper kettle supplied by the Russians by filling it with sable pelts. A decade after the setting up of Yakutsk 150,000 pelts of sable a year passed through its customs house. The pelts of Arctic foxes numbered in the thousands at the same time and those of squirrels in the hundreds of thousands. For a time, Yakutsk became the Tobolsk of the east. By 1650, it was the home of about three hundred thousand Russians. It became the point of departure for tribute-gathering expeditions that carried the tsar's standard to Siberia's remote northeast, into the Amur valley and more important, to the edge of the Pacific.

Just seven years after Beketov founded Yakutsk, the Cossack captain Ivan Moskvitin led twenty men up the Lena's eastern tributaries. He portaged across the Dzhugdzhur Mountains to the headwaters of the Ulia river. After a five-day voyage down the Ulia, Moskvitin's tiny force reached the Sea of Okhotsk. Here they built the wintering place which became the town of Okhotsk. Two years later, Moskvitin and his men returned to Yakutsk carrying the pelts of 440 prime sable and 323 black foxes, a treasure well worth the hardships they had endured. More than four thousand miles east of Moscow and just fifty-seven years after Yermak's victory, the Russians now had their foothold on the Pacific.

The Russians had reached the Pacific. Their supply lines were stretched to the limit from European Russia. Hereafter, it was not just exploration, it was a question of survival in an inhospitable terrain. Siberian summers were short and the winters too cold. Rye, wheat and barley failed to ripen. Fruits and vegetables were difficult to grow. Nor were there herds of bison to feed the multitude. Agriculture was impossible. The tundra was home to herds of reindeer, Eskimos and nomads. The taiga had forests. It sustained hunters and trappers. Either way, agriculture was impossible in the lands east of the Yenisei.

The question of food concerned every military governor who served east of the Yenisei during the seventeenth century. Since no crops grew in the tundra and the taiga lands to the north and the east, the tsar's governors looked south. Between them and the south lay the Stanovoy and Yablonsky ranges. These two mountain ranges divided eastern Siberia's wild Taiga from the Amur River valley. Beyond them lay the Amur Valley. Perhaps its floodplains could solve the problems of seventeenth century hungry Russians. Hence the vanguard of the Russian conquest of Siberia turned south during the 1640s after half a century of moving east.

Chapter-9

The Amur Valley and Russo-Chinese Relations

AFTER THEY FOUNDED YAKUTSK, THE RUSSIANS learned of a valley beyond the Stanovoy Mountains where grain grew in abundance. This was the trans-Baikal region otherwise known as the Dauria. Hoping that he might find more food for Siberia's hungry Russians the military governor of Yakutsk sent out expeditions to investigate the region. In their search for grain, the Russians were to encounter two native peoples, the Daurians and the Gilyaks. They managed to make enemies of both of them. Subjects of the Manchu emperor, the call for tribute from a distant tsar cut no ice with them. Rather, they called the Russians, devils.

In the June of 1643, Vassili Poyarkov set out with 133 Cossacks and promyshlenniki to find the Amur valley, explore it and learn if natives who lived near it grew grain. He followed the Lena's tributary Aldan southward to its source only to discover that he had made the worst possible choice. No fewer than sixty-four rapids had to be portaged. His men lost precious weeks as they shifted their supplies from boats to packs and back to boats again. The first snows came too early that year. Winter struck with full force before the Cossacks even reached the mountain's crest. Nevertheless, Poyarkov decided to press ahead and risk spending the winter in whatever land lay beyond the mountains.

Poyarkov assigned forty-nine of his men to build a wintering place in the mountains with instructions to join him in the spring. Then he and the rest of his men crossed the Stanovoi passes in December and reached the Zeya River, one of the chief tributaries of the Amur. Poyarkov's force now found itself in the Dauria, a wild but fertile domain that stretched along the Zeya and Amur rivers for several hundred miles until it shaded into the lands of the Goldi and Gilyak fishermen further east. Here grain was the staff of life as it was in Russia. For the first time since they had crossed the Urals, the Russians found people who preferred farming to hunting.

From Poyarkov, the Daurians were to learn, as Siberia's natives to the west already had, that no matter how much they gave or how willingly they gave it, the Russians always wanted more. Their greed was insatiable. They would devastate their villages if the natives failed to provide it. In this case, the Daurians offered fur tribute and food, including the oats and buckwheat the Russians had hoped to find. They insisted that they knew nothing about the silver, lead and copper mines of which the Russians had heard rumors. Poyarkov demanded hostages. Certain that the Daurians were concealing silver, he tortured the people who had fallen into his hands. Thereby, he turned a land of willing tributaries into a nest of dangerous enemies. As the snow deepened and the small game disappeared, the Russians tried to take more food from the now hostile Daurians at gunpoint. When their victims fled and left them with no food at all, Poyarkov's men began to hunt the Daurians like wild game, shooting them and roasting their flesh over open fires. These people remembered the Cossacks' brutality for centuries to come.

A diet of pine bark, roots, pirated supplies, stray creatures of the forests, and the flesh of natives supplemented by that of comrades who died along the way kept fewer than forty of Poyarkov's men alive by the time the remnants of the force he had left in the mountains joined him in the spring of 1644. As they worked their way down the Zeya to the Amur, the Russians found that their reputation had preceded them. All along the river, the natives were on the defensive. They forced the unwanted explorers to fight their way through one ambush after another. For the first time, the raw brutality of the Russians was turning against them.

That fall, the Russians entered the territory of the Gilyaks. Here the Amur broadens to a width of ten miles before flowing into the Sea of Okhotsk. Among the Gilyaks, there were no farmers, only hunters and fisherfolk. The Gilyaks dressed in fish skins warmed by robes of dog fur in the winter. To avoid a second hungry winter, Poyarkov and his men seized Gilyak chieftains and held them hostage. This was to guarantee regular delivery of food. Again they abused their captives unnecessarily, making as many enemies among the Gilyaks as they had among the people of Dauria. With hatred rising in his wake, and only a handful of his command still alive, Poyarkov dared not retrace his steps. He therefore continued to move ahead, destined to cover more than five thousand miles in a huge loop that circled the entire southeastern corner of Siberia before he returned to Yakutsk.

Kept strong by the Gilyak's reluctantly delivered provisions, Poyarkov's Cossacks built boats in the spring of 1645 and made their way to the coast of the Sea of Okhotsk to the Ulia River. Here they spent a third winter in the log shelters, Ivan Moskvitin had built six years earlier. The next spring, they followed Moskvitin's route back to Yakutsk, arriving just a few days after the third anniversary of their departure. Poyarkov received no reward. His rough ways of command had turned some of his men against him. Their complaints convinced the military governor of Yakutsk to send him to Moscow for trial and an unknown fate. At the same time, the governor reported enthusiastically that Poyarkov's campaign had brought the treasury five hundred prime sable pelts and confirmation that grain could be found in Dauria. Certain that a Daurian harvest could feed the hungry Russians east of Yeniseisk, the authorities insisted that Russia must conquer the Dauria.

To conquer the Dauria and the rest of the Amur lands became the task of Yerofei Khabarov. Like many young men in European Russia, Khabarov had succumbed to Siberian fur fever in his late teens. He had left the Stroganov's salt town of Solvychegodsk in the mid-1630s to seek his fortune in the Mangazeya fur trade. There he ranged as far afield as the Taimyr peninsula but found only modest profit. The early 1640s found Khabarov and his brother trapping furs in the Lena valley their thirst for fame still unquenched.

For a while, Khabarov tried to follow the footsteps of the Stroganovs by building a salt factory near Yakutsk. When the government took it away from him, he turned to farming. He tried to use native works and transported criminals from Russia to grow grain in the hostile Yakutsk climate. Among his grain fields, he built a flour mill. None of these enterprises satisfied the longings that had drawn him to Siberia's eastern depths. Knowing that fame and fortune awaited the man who conquered the grain fields of the Dauria, Khabarov offered to explore them at his own expense. Not surprisingly, the military governor of Yakutsk gave him his blessing. Since the Russians now knew that the Olekma River offered the best way into the mountains, Khabarov led a force of seventy volunteers along that route in the spring of 1649.

Crossing the Yablonsky rather than the Stanovoi peaks farther east, Khabarov entered the Dauria in the beginning of 1650. He found the countryside deserted, its villages abandoned in haste, and the native's stores of provisions hidden or burned. During a brief meeting with three Daurian princes, he learned through interpreters that their people had fled for fear that he would renew Poyarkov's reign of terror. Instead of trying to win allies for Russia, he did little to ease their fears. Seeing only frightened natives around him, Khabarov anticipated an easy conquest. A reign of terror still seemed to be the easiest way for Siberia's vastly outnumbered conquerors to keep its people firmly under control, and to "scoop everything out."

Then Khabarov learned that a complex system of alliances tied the

Daurians to Manchuria and thus to the new Manchu lords of China. Uneasy about risking a war with the Chinese, Khabarov hurried back across the mountains at the end of May 1650 to report back to the military governor of Yakutsk. Emphasizing the difference between the fertile lands of the Amur valley and the barrenness of eastern Siberia, Khabarov urged his superior to let take the Amur for Russia. "This land of Dauria will be more profitable than the Lena," he promised, "and, in contrast to all of Siberia, will be a region bountiful and abundant." He added, "in the great river Amur there are more fish than in the Volga." There was enough grain in the Amur valley, Khabarov asserted, to feed twenty thousand Russians every year. It could be sent to Yakutsk in less than two months rather than the four years needed to ship flour from Verkhoture. But he also warned than China might be a formidable enemy if she chose to defend her vassals on the Amur. He estimated that Russia would need to send at least six thousand soldiers armed with muskets and cannon to hold the Dauria and the Amur lands to the east. Since all garrisons in Siberia numbered fewer than half that figure, an army would have to be sent from Moscow to be ready if China intervened.

While he waited for the tsar's reply, the military governor decided to act on his own. If he did nothing, and the tsar supported the plan to take the Dauria, he might be punished for not acting promptly. Given the choices, the tsar's chief official in Yakutsk gave Khabarov permission to assemble another force of volunteers to which he contributed weapons and a unit of twenty-one men. With fresh horses, several small cannon, and nearly twice the men he had commanded the year before, Khabarov crossed the mountains again in the fall of 1650. This time, the Daurians refused to pay tribute and defended their land in several battles. Cossack muskets and sled-mounted cannon were pitted against native bows and arrows. After reinforcements had increased his force to over two hundred at the beginning of June 1651, Khabarov followed Poyarkov's route down the Amur valley. He intended to establish a line of frontier forts to cement the Russian's hold upon it, just as his predecessors had done in the basins of the Ob, Yenisei and Lena. Often, the natives fled. When they did not, the Russians battered down their defenses with cannon fire before butchering the men, raping the women and driving off whatever livestock they found. Over the next decade, they would ravage the Dauria so thoroughly that its people could no longer produce the grain for which the Russians had come in the first place.

By the middle of September, Khabarov's men had reached the mouth of the Sungari River, one of the Amur's tributaries. Two weeks later, they built a fort in the vicinity of present-day Komsomolsk-on-Amur and called it Achansk. Again the natives offered fur tribute and provisions. The Russians responded by demanding more and torturing hostages. The natives rose against them. Even though Khabarov's men killed nine hundred natives and suffered but one casualty in crushing the uprising which followed, the victory cost them dearly in other ways. Convinced that they could not drive the Russians out of their lands, the people of the Amur valley turned to the Chinese. While the Russians wintered in Achansk, a well-armed Chinese army summoned by the natives was marching toward it.

A complex series of events in Moscow, Peking and Yakutsk broadened the conflict on the Amur during 1651 and 1652. Khabarov's first winter in the Dauria convinced him that Russia must colonize the Amur lands not just conquer them. He urged men and women condemned to Siberian exile be transported there to help the land yield the full measure of its fruits. In response to Khabarov's proposals, the tsar had decided to send an army of three thousand men under Prince Lobanov-Rostovski, one of his best commanders. At the same time, with an absurd lack of understanding of Asian affairs, the military governor invited the Chinese emperor to become a Russian subject. Sensing that the Russians were serious about claiming the Amur lands, and now freed from civil wars, the Chinese moved against them in 1652.

On March 24, 1652, two thousand Chinese attacked Achansk and broke open its defenses with artillery. Here was an enemy against which the Russians were not prepared to fight. Even though most of the Chinese fought with bows and arrows, their cannon was superior and the few fire-arms they carried surpassed the Russian's guns in firepower. Learning that the Chinese had orders to take them alive, Khabarov and his men managed to drive them off with moderate losses. Their victory reminded them that they faced dangers they faced dangers they still could not measure. How many men could the Chinese put into the field and how quickly? Had their main army attacked Achansk, or had the Russians only faced a vanguard force? How well-equipped were the Chinese? The Russians had many questions and few answers.

With fear beginning to rule his calculations, Khabarov turned back up the Amur. He plotted his march along the Yablonsky Mountains to avoid the Chinese. At one point, a force of six thousand Chinese got ahead of him, but good fortune helped his men slip past the enemy under cover of darkness. Two hundred miles farther on, the Russians met 117 Cossacks sent from Yakutsk to reinforce them more than a year before. These turned out to be more interested in plundering the natives than in fighting the Chinese. When some of Khabarov's men joined the Cossacks in a mutiny, he had to spend the rest of the summer chasing them down and trying to avoid the Chinese at the same time. Catching up with the mutineers in mid-October, Khabarov stormed their hastily built fortifications and flogged to death those who survived his attack. With tensions of the mutiny still festering, the Russians faced a fourth winter in the Amur lands.

While Khabarov waited for new instructions, the authorities in Moscow sent a detachment of 150 men under Dmitri Zinoviev to strengthen his small force. When the two detachments finally met in the fall of 1653, Zinoviev demanded full command. Khabarov insisted on seeing the orders that allowed him to claim that authority. Zinoviev placed him under arrest and sent a formal complaint to Moscow. Khabarov's men protested. Unable to win their loyalty and not willing to remain in the Amur valley without it, Zinoviev returned to Moscow taking Khabarov with him.

Deprived of his property and rank, Khabarov made the fifteen-month journey to Moscow. He waited for the better part of a year before being acquitted of Zinoviev's charges. Eventually he was given a minor rank on condition he return to Siberia. "Because of my poverty," he wrote, "I am perishing from hunger." Then he fell from sight. The only other mention of Khabarov comes from the records in 1658 when the Siberian Office issued orders that he be put in irons if he refused to serve as a guide for an expedition on its way to the Amur. After that the account breaks off. We only know that he lived for ten more years. We do not know when this poorly rewarded conqueror died or where.

Khabarov's victories left a dangerous and uncertain legacy on the Amur. Following in Poyarkov's footsteps, he had found the grain needed to feed the Russians in Siberia's eastern lands. It would take centuries to undo the hatred that his brutality had stirred up among the natives. At the same time, the stories of wealth that his men had spread had infected the Russians in the Lena valley with what one historian has called an epidemic of "colonizing fever" that drew them away from Yakutsk into the Amur lands. By 1653, so many Russians had pulled up stakes and struck out across the mountains to plunder the lands of the Dauria, that the valley of the Lena was becoming depopulated. The damage they inflicted as they tried to claim tribute from the Daurians wiped away the Russian's last chances for provisioning eastern Siberia from the nearby Amur valley grain fields.

The outbreak of the First Northern war between Russia and Poland obliged the tsar to recall the regular forces he had sent to Siberia. Only a small band of irregulars remained on the Amur to defend the flood of Russian immigrants against the Chinese. These men were no match for the Chinese who killed or captured them all before 1658. Then having wiped out all the Russians on the Amur below Nerchinsk, the Chinese withdrew, leaving Dauria to become a haven for outlaws and renegade Cossacks for the next fifteen years. When the Russians returned to the Amur lands in 1672, they again treated its handful of Chinese and Manchurian hunters and traders with such brutality that China's rulers decided to go to war again in the 1680s.

The Russo-Chinese conflict of the 1680s saw the Chinese take the frontier fort that the Russians had rebuilt in Albazin in 1685. They then withdrew, only to have the Russians return and rebuild more modern defenses a few months later. When the Chinese mounted a new siege in July1686, Albazin had more than eight hundred defenders supplied with enough food and water to last for a year. It also had a dozen cannon and more than two tons of gunpowder. This time, the Chinese were prepared to bring Albazin to its knees. After a four-month siege, the Russians asked for a truce. Signed in September 1689, the Treaty of Nerchinsk forced the Russians to destroy Albazin and renounce all claims to the Amur lands. For the time being, the Amur had been closed and the hope of using its grain to feed the Russians in Siberia's eastern lands brought to an end.

Conclusion: With Siberia's southern boundaries fixed at the crest of the Stanovoi and Yablonsky mountains till the middle of the nineteenth century, the Russians turned towards Siberia's northeast. Eventually their quest for furs and food would carry them beyond the Bering Strait and the North Pacific to North America. In the process, they unwittingly solved one of the greatest puzzles facing the geographers of the seventeenth century.

Chapter-10

Finding a Passage between Two Continents

FINDING A SEA ROUTE BECAME IMPORTANT for trade when the limitations of what could be carried on a camel's back came to be realized. Silk and spices brought on camel's backs across the Silk Route commanded astronomical prices for the most microscopic of quantities. It was the dream of merchants to bring large amounts of precious commodities from the Orient quickly and with less hazard. Visions of finding a sea route to Cathay fired the imaginations of navigator's the minute the mariner's compass was discovered. They could now sail beyond sight of land. If trade could be shifted from the backs of camels to hundreds of times larger hold of ships, the prospects for profit were immense. Such visions drew the Portuguese around Africa's Cape of Good Hope and across the Indian Ocean just as they guided Columbus's voyage to the New World.

Reaching the Orient through sea was thus the aim of navigators. The Spanish and the Portuguese found that North and South America stood between them and Asia. Therefore, they set about finding a waterway that would carry them through. Magellan found one such passage eight thousand miles from Europe near South America's southern tip. Its waters however were too perilous for none but the most skillful and daring of pilots. The explorers of England and Holland therefore turned north. They hoped to find a less arduous passage close to their home ports. To England's Martin Frobisher, Henry Hudson and William Baffin, a Northwest Passage through the lands of North America seemed the most likely way to reach east. At the same time, Sir Hugh Willoughby, Richard Chancellor and William Barents searched for a Northeast passage that would carry them north of Eurasia to China.

On a voyage that became the high-water mark of the search for a Northeast Passage, Barents and his crew spent the winter of 1596-1597 on the ice-locked island of Novaia Zemlia, which separates the Barents and Kara seas almost directly north of the Ural's northern tip. Like his predecessors, Barents failed but each failure enlarged navigator's knowledge of the world's geography. They kept alive the hope that another path around the continents may yet be found. As Europeans continued their search, the most vital of all questions remained unanswered: Were Asia and North America actually separate continents or were they joined somewhere north of China? Even if the waters above North America or Eurasia could be navigated, would it be possible to sail south to China or would land block the way?

Russian exploration of the north proceeded at its own pace and along lines thought up by the Russians themselves. They were cut off from the learned men of Europe during the Age of Discovery and unaware of the questions that commanded their attention. They explored Siberia alone and in their own way. They seem to have been unaware of the mariner's compass till the end of the seventeenth century. Like travelers of yore they took their bearings from the sun and stars. They had none of the accumulated wisdom of the West to guide them. They could only sail with the wind not against it. Nonetheless they worked their way along the north of Eurasia's coast from their settlements on the White Sea to those on the Ob and the Yenisei. Sailing their small coastal vessels east from Arkhangelsk, Russian traders had reached the Ob Gulf, the Taz river and Mangazeya by 1600. Just a few years later they made their way to the mouth of the Yenisei and sailed up it to Turukhansk.

Fearing English and Dutch penetration into Siberia, Russia closed the Mangazeya seaway in 1619. Pomor activity in Northern Asia declined and the bulk of exploration in the 17th century was carried out by Siberian Cossacks, sailing from one river mouth to another in their Arctic-worthy *kochs*. The Taimyr peninsula, the northernmost point of Eurasia barred their way further east. A number of attempts to sail from the Yenisei to the Lena via the Kara and Laptev seas on the peninsula's western and eastern sides failed. Filled with mountains of pack ice the Arctic remained impassable.

Unable to round the Taimyr peninsula, the Russians moved on to the Lena river by a combination of river routes and portages and then resumed their sea voyages from its mouth. Faced by some of the continent's most hostile natives, sharp mountain crags, rivers whose tributaries no longer came close to connecting and four mountain ranges all of which stood above the Arctic Circle, the Russians considered conquering Siberia's northeast from the sea rather than fighting their way overland. From one to twelve miles wide, the mighty Lena formed a majestic entrance to the new lands they hoped to explore. If the seas east of its mouth could be navigated, it would be relatively simple to ascend the rivers of Siberia's northeast and claim the lands around them. Some of the Cossacks and *promyshlenniki* at Yakutsk were willing to try. Others preferred to face the howling storms that lashed the steep face of the Verkhoyansk Mountains and work their way into the northeast by land rather than venture into the ice-filled seas of the Arctic.

In the late 1630s, Cossacks and *promyshlenniki* set out to pit their flat-bottomed wooden boats-built without tar or nails and with rigging and cordage made of deerskin strips attached to stiff deer hide sails against the perils of the Arctic. Able only to run before the wind, such craft bore no relation to the triple-masted vessels that seventeenth century European explorers were sailing across the world's other oceans. Unsteady on even the tamest seas and with none of the nautical instruments that Europeans used to chart a course, the boats that Russia's Cossack and promyshlenniki Vikings sailed from cape to cape and bay to bay east of the Lena's ice-clogged delta seemed to be in every way unequal to the perils that lay ahead. Shifting mases of ice could crush whatever lay between them in an instant, and the jagged edges of pack ice could cut the twigs and thongs that held a boat's siding in place with predictable ease. Many of the men who entrusted their lives to these fragile vessels never returned but a few became famous. Among them Mikhail Stadukhin and Semyon Dezhnev who explored both land and sea routes into Siberia's northeast in the 1640s became noted explorers and bitter rivals.

Stadukhin and Dezhnev had grown up on European Russia's White Sea coast. In the 1630s, infected with fur fever, both men made their way to Yakutsk. Stadukhin became wealthy and won the friendship of the military governor. By 1640, the first expeditions to the lands east of Yakutsk had discovered the Jana and the Indigirka, the first rivers east of the Lena. Returning with a wealth of prime sable pelts and stories of other, greater rivers yet to be discovered, the men who crossed the first barriers of Siberia's northeast spoke of the fabled "Pogycha" river along whose banks walrus ivory was said to piled in heaps and coal-black sable ran in herds. According to other reports, there was a lake filled with pearls not far from the "Pogycha" and a mountain nearby from which silver could be obtained by firing arrows into it and picking up fragments they knocked loose. The prospect of silver mountains and black sable herds was irresistible. Having already been on earlier expeditions into the borderlands of the northeast, Stadukhin and Dezhnev threw themselves into the search for "Pogycha".

Stadukhin's wealth and friendship with the military governor of Yakutsk won him command of an expedition sent out in the spring of 1642 on a forty-four-month overland journey to the Indigirka river valley. After spending the winter in temperatures that fell to -70 degree centigrade and losing nearly all their horses in a battle against the natives, Stadukhin and his men built boats on the Indigirka, sailed down it to what they called the Freezing Sea and made their way east for about three hundred miles to the Kolyma River. There they laid the foundations of Nizhnekolymsk, the frontier fort that would mark the eastern edge of Siberia's fur frontier for the rest of the decade. In 1646, they returned to Yakutsk with a fortune in able pelts. Based on what he had heard from native she had captured, Stadukhin hastened to report to the military governor that the famed "Pogycha" lay a scant three day's sail east of the Kolyma's mouth.

Stadukhin's report unleashed a rush of trappers and traders into the Kolyma basin. Believing that the "Pogycha" could be reached more easily by sea, more than four hundred Russians applied to the authorities in Yakutsk for permission "to sail down the Lena and to travel by sea to the Indigirka and Kolyma rivers for trade and fur collection." Those who were well-connected with the governor and his circle now looked to Stadukhin to lead the way to the "Pogycha". Some of the newly arrived agents sent by some of Moscow's richest merchants to search out new openings on the Siberian frontier turned to Dezhnev, whose exploits in collecting fur tribute from the warlike Yukaghir had won him the reputation of a man who could get things done. They hoped that Dezhnev's talent for turning adversity to advantage might yield rich returns.

With Dezhnev setting out from Nizhnekolymsk on the Kolyma and Stadukhin sailing from a base on the Lena, the search for the "Pogycha" began in real earnest during the summer of 1647. After unusually thick ice floes prevented either expedition from entering the Freezing Sea that summer, both spent the winter and spring making new preparations so as to be ready to try again in 1648. This time, Dezhnev had better luck. Setting out with ninety men in seven boats, he was now destined to lead one of the scant handful of Russian sea voyages to make history. Although a storm wrecked two of their craft on the Kolyma's mouth, the rest of Dezhnev's force entered the Freezing Sea without incident and began their voyage along the coast. Because the Kolyma was the easternmost Asia river to flow into the Arctic, they never found the "Pogycha" River. Instead they reached the northeastern tip of Siberia at the beginning of September, rounded the Chukotka peninsula, and passed the east cape (now called Cape Dezhnev), thus proving that North America and Asia were separate continents.

Ignorant of geography and unaware of the larger questions that commanded cartographer's attention in the West, Dezhnev and his comrades had no sense of the importance of what they had done. In retrospect, their success is sailing their crude craft for more than a thousand miles through the East Siberian and Chukchi seas seems all but impossible. As recently as two decades ago, some experts still were challenging the claim that Dezhnev's men had done so. The distance was too great, they insisted, the men's experience too limited and their knowledge of navigation nonexistent. Dezhnev's boats were too fragile and their motive power too weak. Yet Russian scholars have now proved beyond a doubt that the events that Dezhnev recounted in a disposition taken down several years later actually happened.

About the time that Dezhnev passed the cape that now bears his name, the luck that carried him around Asia's northeastern tip turned sour. A storm sank all but one of his boats, and only twenty-four of his men managed to reach the desolate southern coast of the Chukhotka peninsula. "all of us took to the hills," the almost certainly illiterate Dezhnev explained in his disposition to the military governor of Irkutsk. "We were cold and hungry and without shelter and barefoot and I, poor Semeika and my comrades went to the Anadyr in just ten weeks, reaching the lower reaches of that Anadyr River near where it flows into the sea." Now nearly six thousand miles east of Moscow, a full twelve hundred miles northeast of Yakutsk, and nearly half a world east of Greenwich, Dezhnev and his men did not know where they were nor how they could return. Their boats were gone. Marooned in the northeastern most corner of Asia with only the sun and stars to tell their location, they could only guess that Yakutsk lay far away to the southwest. Unable to find food or wood, Dezhnev divided his men into two groups to search for a better wintering place. One group disappeared, but Dezhnev and eleven companions survived to claim Siberia's northeastern most corner for Russia.

In his disposition Dezhnev explains, "Going by sea from the Kolyma River to the Anadyr River one comes to a cape that extends far out," he explained. From that point, he had sailed on to the mouth of the Anadyr River on the Bering Sea, thereby proving to the careful cartographers of Europe that Asia and North America were separate continents. But, probably because the military governor of Yakutsk favored Stadukhin as the best man able to enlist new tribute payers for the tsar, Dezhnev's report never reached Moscow and remained buried in the Yakutsk archives for nearly a century.

Far away on the Anadyr River, Dezhnev and his eleven companions tried to act as the tsar's agents. They collected fur tribute from the natives, trapped sable on their own, and collected walrus and mammoth ivory when they found it. In the summer of 1649, they worked their way some 350 miles west to the Anadyr's headwaters and built a new wintering place. Now in desolate lands where trapping was poor, they still did not know how to return to Yakutsk. For a while, Dezhnev seems to have toyed with the idea of building new boats and trying to return the way he had come. But the currents of the Bering Sea flowed in the wrong direction. A clumsy Cossack boat, tied together with thongs and willow twigs and powered by oars supplemented by a fixed deerskin sail, could never have taken the reverse course.

A small detachment from Nizhnekolymsk suddenly appeared at their camp the next spring. Their rescuer turned out to be Stadukhin who insisted that his higher rank entitled him to full command of their combined force. Dezhnev now had to become a subordinate of the man whom he had just defeated in the race to claim Siberia's northeast.

But Stadukhin had not been alone in searching for a way across the mountains. For most of the way from Nizhnekolymsk, he had followed the trail of a smaller expedition led by Semen Motora, a Cossack officer who had spent the previous two years searching for a passage to the coast. He had overtaken Motora just days before his forces met Dezhnev's. Thus it had been Motora who had done the work but Stadukhin who claimed the credit for finding a way across Siberia's last mountain barrier. For the next eight months, these three proud, self-centered men feuded and fought. With Stadukhin refusing to relinquish command, several of his lieutenants defected to Motora and Dezhnev, who eventually joined forces and left Stadukhin to search for richer sources of fur further south.

Now working together, Dezhnev and Motora explored the lands along the Anadyr, finding a fortune in walrus ivory on the sandbar that filled the river's mouth at low tide. In a month, they dug up nearly three tons and then, with summer coming to an end, made their way back to their wintering place in Anadyrsk. The next spring, Motora was killed in a skirmish with the natives, but

Dezhnev continued to collect ivory in the barren lands that spread across Siberia's northeast for most of another decade. Yet he did not remain out of touch with the authorities at Yakutsk as he had in earlier times. Although no one repeated his voyage in the seventeenth century, expeditions from Yakutsk traveled overland to Anadyrsk every year once Motora and Stadukhin had found the way. It was a literate officer in one of these expeditions who took down the disposition that contained the details of Dezhnev's historic voyage.

If the broader significance of Dezhnev's discovery remained unknown for a century, his entry into the Chukhotka peninsula brought the Russians into contact with natives more warlike and better able to resist than any they had yet encountered. Unlike the weaker and smaller tribes to their west, who had paid fur tribute to their stronger neighbors for centuries, the proud Koryaks whose settlements stretched south from the Anadyr to the Kamchatka peninsula, were determined to resist the degradation. When the Russians tried to weaken them by provoking conflicts with their rivals, as they had done when they pitted the Ostyaks against Samoyeds, Tungus against Buryats and Yukaghir against Chukchi, the Koryaks quickly forged alliances with other tribes. Determined to defend their lands, the Koryaks used captured firearms along with the most effective of their native weapons to keep the Russians north of the Anadyr River until the very end of the century. Only then could the Russians manage to enter the Kamchatka peninsula, the last remaining reservoir of rare pelts in Eurasia.

Chapter-11

The Demidovs and Siberian Iron

AT THE END OF THE SEVENTEENTH CENTURY, the statesmen and sovereigns who ruled Europe believed that only those nations possessing unlimited access to natural resources and the technology to manufacture them could become rich and strong. This belief had led England, Holland, France and Spain to build colonial empires. Always on search of new empires, the merchant adventurers of England and Holland viewed Russia in similar terms. Russia was to them a rude and barbarous kingdom. It held a fortune in furs as well as the grain Europe needed to feed its growing cities. Russia also had hemp, tar and timber in abundance, all essential for modern navies. For these reasons, Europeans had been reaching into Russia since the days of Ivan the Terrible. By the middle of the seventeenth century, the English had even built a ropewalk on North Russia's White Sea coast to manufacture cordage for their fleet.

The Russians therefore stood in danger of being exploited by nations with greater access to wealth and technology. "When natives sell their products in a raw state," one Russian statesman warned, "it is the inhabitants of other countries who work up the raw materials and receive the great return on their labors while the former possessors receive only a meager sustenance." Aware that the richer and more advanced nations of Western Europe would devour Russia if it was not their equal, Peter the Great realized that there was no time to lose. Russia needed to build an army equal to any in the world. To do that it would need to build its own arms.

Peter the Great took Russia's helm in 1689 with a clear sense of urgency and purpose. He understood that Russian armorers had to make Russia's weapons from Russian iron and steel if Russia hoped to stand among the nations of the West. Russia had no iron industry aside from a few small foundries and a handful of armories in Moscow and Tula. All of Peter's predecessors had ordered every military governor they sent across the Urals to look for iron ore in Siberia. Some had tried to enlarge Russia's iron industry by hiring foreign experts to work with imported iron. None of these efforts had succeeded. Small deposits of iron ore had been found near the Stroganov's saltworks at Solvychegodsk and a few other places. At the end of the seventeenth century, most Russian iron still came from bogs, dug by peasants, heated in small smithy forges and then hammered into small bars or strips. For higher quality iron and steel, the Russians had to rely upon Sweden, the very nation she would have to fight to gain access to the Baltic.

The discovery of high-quality iron ore at Nevyansk near Siberia's western frontier transformed Russia's iron industry in scarcely more than a decade after Peter the Great came to the throne. Set in the midst of a large network of rivers and surrounded by forests that promised abundant supplies of fuel, this ore was pronounced by experts "better than that found in Sweden." Now, foundries had to be built, a labor force assembled in the wilderness and master workmen trained to work with iron. The scale of operations was immense. Peter first assigned this task to the governor of Verkhoture. He failed in this task even as Russia went to war with Sweden. Tired of the "botched iron" and "wry-mouth cannon" that the new foundry at Nevyansk sent his armies and arsenals, the tsar turned it over to Nikita Demidych Demidov, an ironmaster and weapons maker who had won his confidence.

Demidov's powerful sloping shoulders bespoke a life spent at the forge. Some accounts claim that he first won Peter's favor by making for one of his favorites a perfect copy of an intricately designed pistol that had come from the armory of one of Central Europe's most famous gunsmiths. Others insist that he rose above Tula's other arms makers by producing modern flintlock muskets to Peter's specifications at short notice. All we know for certain is that the tsar stopped at Tula on one of several journeys he made between Moscow and the front during his first war with the Ottoman empire and that sometime in late 1695 or early 1696 Demidov presented him with six firearms of his own design. Impressed by Demidov's understanding of modern weapons technology and his ability to manage men, Peter made him a promise. "Strive to expand your factory," he told him then. "Strive to expand your factory and I will not abandon you."

For several more years Nikita Demidov worked at Tula. He became the first ironmaster there to use water power, and after buying up several other foundries, he began to produce flintlock muskets that met the government's specifications at a fraction of what it cost to import them from Europe. With the beginning of the Great Northern War between Russia and Sweden in 1700, government orders began to pour in. In less than a year, Demidov had contracts for twenty thousand flintlock muskets. Peter, it seemed, had already kept his promise that he would "not abandon" him. Then his patronage took a different form.

In 1702, Peter the Great, anxious to turn the failed foundry at Nevyansk over to private hands in return for pledges of cheap iron and well-made cannon offered it to the artful Nikita. Demidov seized the chance, sensing the opportunity to build that on a scale that he could never have achieved at Tula. At Nevyansk and in the ten thousand square miles around it, Nikita and his son began to build an empire, the base of which from the Chusovaya River could carry barge loads of iron and weapons into the heart of European Russia. The Chusovaya, and, on the Siberian side, the Tagil and Neiva would eventually bind a sprawling network of mines, timbering camps, forges and foundries that would move men, metal, weapons, and supplies back and forth across the mountains.

To make iron and guns for the tsar, the Demidovs needed not only ore that was plentiful but labor that was cheap. Nikita therefore took the unprecedented step of asking the tsar to let him buy serfs to work in his foundries for the only sources of local labor around Nevyansk were a handful of Russian peasants and a few scattered native settlements. Only the tsar, the Royal Treasury and the nobility had the right to own serfs in Russia. Peter's readiness to grant that privilege to a commoner indicates how urgently he wanted an iron and weapons industry that did not depend on metal or technicians from the West. With the tsar's support, Demidov became the first commoner to own serfs in modern Russia. By doing so, he made certain that some of Russia's best foundry masters, metalworkers and gunsmiths in the eighteenth century would be serfs.

The move to Nevyansk in 1702 cemented an alliance between the Demidovs and the Romanovs for the next half century. Demidov iron became the first of Siberia's products to be exempted from the duty imposed on goods crossing the Urals into Russia. By 1720, the Demidov forges and foundries had become the exclusive suppliers of iron to the Russian navy. Obliged to organize large-scale logging operations to fuel their forges and foundries, the Demidovs began to cut timber for government shipyards. The most successful frontier entrepreneurs in Russia since the Stroganovs, they gave Peter and his successors the iron industry Russia needed to win a place among the great powers of the West. By the time, Russia became the world's leading producer of pig iron in the 1780s, two out of every five tons of it came from the foundries that Nikita and his descendants had built in Siberia.

As tsar and blacksmith traded gifts and favors in the early 1720s, both were approaching the end of their reigns, for both would die in 1725. With weapons forged from Demidov iron, Peter's soldiers had won a foothold on the Baltic and carved a place for Russia among the great powers of Europe. Demidov iron had also anchored and braced Russia's ships and served a hundred other vital purposes in building Peter the Great's new empire. The same iron had also built a kingdom of forges and foundries, for Nikita added six more ironworks to the one at Nevyansk before he died. Each of these increased his already unmatched ability to produce more of the metal that made his fortune. In 1725, he therefore left behind the beginnings of an empire that his son would expand to dwarf his achievement. By that time, Demidov forges were producing more than ten thousand tons of iron a year, or more than half of all the iron produced in the Russian Empire.

As with the second generation of so many of the dynasties that led Europe and America into the Industrial Revolution, Nikita Demidov's eldest son, Akinfi, built upon his father's achievement according to his genius and the laws of mathematical progression. Born in 1678, Akinfi learned to make iron and steel at Freiburg, one of the greatest centers of metallurgy in Europe. He then returned to Russia to work with his father just before they took over the Nevyansk foundry. For more than two decades, the two men labored to build their empire. They did so on the basis of Nikita's old-fashioned intuition and experience, not Akinfi's knowledge of modern science. Nikita remained a patriarch who ruled his enterprises and his son with equal sternness, and, although he had the good sense to allow him some latitude, much of what Akinfi had learned in Freiburg had to be set aside to do his father's bidding.

When Nikita's death freed his hands, Akinfi combined his technical training with the stubbornness, confidence and entrepreneurial skill he had inherited from his father to expand the Demidov realm of forges and foundries in the Urals into a Siberian industrial empire. Even more than his father, Akinfi's life revolved around the forging and casting of metal, and each of his enterprises he built in Siberia's western lands received devoted personal attention. Making metal was his passion, and no detail was too small or any problem too large to consider. "Just like small children," he once wrote, "factories demand careful supervision." During the twenty-one years that separated his father's death from his own, Akinfi made certain that his enterprises were the best-watched industrial progeny in Russia.

Akinfi's genius for forging metal and managing metal works allowed him to surpass all rivals in producing iron at lower cost and higher profit. Using only a fourth as much labor, his foundries yielded twice the output of their government managed counterparts. By building better sluices and more efficient blast furnaces, he produced iron and sold it to Moscow and St. Petersburg at more than ten times the profit earned by foundries run by the government. At the same time, he towered over his handful of private competitors. No other Russian iron maker produced more than a sixth of what the Demidov foundries did during the first half of the eighteenth century, nor could any other iron forged in Russia match Demidov quality. Russia could not even begin to consume the output of Akinfi's foundries. By the time he died in 1745, nearly seven out of every ten tons of metal that came from his ironworks were being sold abroad. From a nation obliged to import iron at high cost from potential enemies in 1700, the Demidovs had transformed Russia into one of the world's largest iron exporters in less than fifty years. The English, who, in the days of Ivan the Terrible had tried to melt and smelt iron in Russia themselves had now become Russia's willing customers. "The iron is pure, even in width and girth, and of highest quality," one enthusiastic English buyer wrote in the 1730s. Especially at the price the Demidovs sold it, it was well worth paying the cost to ship it to England.

Not satisfied to be Russia's largest producer of iron, Akinfi began to branch out into other areas. Like iron, copper had been in short supply throughout Russia's history. Every tsar since Ivan the Terrible had commissioned special prospectors and foreign experts to search for it in the far north and Siberia. In the sixteenth and seventeenth centuries, the handful of finds, whose meager yields of low-grade ore had played out quickly, underlined the importance of location richer lodes. Intrigued by tales of mines deep in Siberia that had worked long before the Russian's coming and looking for new challenges to test his skill, Akinfi became Russia's largest smelter of copper. More expert in metallurgy that the men who had gone before him, he shifted the focus of Russia's search for copper away from the north to the Altai foothills near the frontier of Mongolia in Siberia's south. There, more than a thousand miles south and east of the Urals, he found the treasure in copper ore that prospectors had been seeking since the days of Yermak.

In 1724, after long months of searching, Akinfi's prospectors uncovered ancient forges in the valley of the Shulba River and then, on Christmas Day 1725, found the mines near the shores of Lake Kolyvan that once had supplied them. In the spring of 1726, Catherine I granted him permission to mine copper in the Altai lands. Within a decade, her successor, the Empress Anna, forbade all others from mining or smelting that metal anywhere in the vast territories between the Urals and the Ob. Still, no matter how talented or powerful, no single man could exploit Siberia's mineral wealth to the fullest. Nor would the forces that shaped Russian politics permit so much power and wealth to be concentrated in the hands of one family. Others therefore had to be given a share of the opportunity, even though none could hope to rival the Demidovs until the heirs of Nikita and Akinfi shifted their interests from Siberia's metals to waiting upon Russia's sovereigns in St. Petersburg at the end of the century.

Unlike their father and grandfather, who had served a tsar seeking to win for Russia a prominent place among the great powers of the modern world, the Demidovs who left iron making and copper smelting to become diplomats, statesmen, and aristocrats in St. Petersburg served empresses and emperors who ruled the world's largest contiguous empire. As one of Europe's great powers, this new empire formed an integral part of the alliance system that divided and united Europe at various times in the eighteenth century. Its armies fought in Europe's wars as well as its own. Cossack squadrons in the outskirts of Berlin and the end of the Seven Years' War in 1761 and the parade of Russian divisions down the Champs-Elysees in 1814 provided the most dramatic evidence of Peter's achievement in making Russia a European power. Other factors tied the empire he had created to Europe as well. Like the rest of Europe in the eighteenth century, the Russians devoted themselves to slavish imitations of the Versailles court, and the growing English market for Demidov iron became only one of many instances of Russia's growing involvement in the trade and commerce of Europe.

Trade and the quest for new technology had begun to draw Russia into the world of European learning even in Tsar Peter's time. As more Russians followed the path of Akinfi Demidov in studying in the West, Western science came to Russia along with European scientists. In search of new and fertile areas of scholarly inquiry and scientific exploration, these disciples of Western learning turned to Siberia. There, botanists, historians, geologists, and above all, geographers found virgin scientific ground, all of it untouched by the hand of modern science and scholarship.

Chapter-12

Bering's First Voyage

IN HIS PASSION FOR WESTERN SCIENCE AND TECHNOLOGY, Peter the Great built an academy of sciences that became the outpost of the scientific revolution in Russia. All of its seventeen fellows named in 1725 had to be imported from Central Europe because no Russians at the time were sufficiently learned to hold an appointment, and it became their task to educate Russia's first scholars. While they taught the Russians, these Europeans studied the natural history and ethnography of the land to which they had come, but geography held their attention most of all. To them it seemed that the time had come to reopen the search for a Northeast Passage.

For men anxious to know if Asia and North America were joined at the top of the globe, the mystery of northeastern Siberia's so-called "glacial cape" had to be solved. "A large strip of land extends far northward toward the so-called but as yet unknown glacial cape," the great German scholar Gottfried Leibnitz explained to one of Peter's advisers. "It is necessary to determine if this cape exists," he continued," [and to learn] if this strip ends at the cape [or converges with America]." Dezhnev's long-ago voyage had answered Leibnitz's question, but not even Russia's powerful tsar knew about the long-forgotten account of his discovery, which had been gathering dust in the Yakutsk archives for the past seventy years. What guided Peter (but remained unknown to Leibnitz and most of his contemporaries) was a later account prepared by Vladimir Atlasov, the explorer who had led the Russians into the Kamchatka peninsula at the very end of the seventeenth century. "Between the Kolyma and Anadyr Rivers there is an impassable 'nose' or peninsula that extends into the sea," Atlasov had written in 1698." Along the left side of this 'nose'," he continued, "there is ice in the sea during summer; while in the winter this same area is frozen over." Ice, not an isthmus connecting Asia with North America made the waters around Siberia's East Cape impassable, the natives had told Atlasov. While Leibnitz and his fellow scholars continued to ponder the land barriers that might stand in the way of a Northeast Passage through the Arctic, Peter the Great and his advisers contemplated how its waters might be charted and a pathway found through the ice.

Unlike the European scholars with their urgings for new explorations who continued to pour into St. Petersburg during the last decade of his reign, Peter had other concerns about the waters that stretched east from Siberia's East Cape. By the beginning of the eighteenth century, the Russian's rape of Siberia's fur lands had led to a sharp decline in the number of fur-bearing animals all across northern Asia. Just as Peter's wars placed an immense new strain upon Russia's national treasury, the income from the Siberian fur trade began to decline. With no lands left to claim in Siberia and no new sources elsewhere that could compensate for the shortage of pelts that the Russian's greed had caused, Peter began to look beyond the sea to the fur lands of North America. Farther south, the Spanish and British had already begun to reap rich fur harvests along the coasts of California and Canada. There was every reason to expect that the unexplored North American lands that lay across the Pacific from Siberia's eastern coast could yield similar wealth. In the 1720s, these remained undiscovered. Peter and his advisers felt certain that North America's western lands must extend as far north as Europe and Asia. Peter was fearful that Spain and Britain might well try to prevent Russia from staking a claim in the New World should they sense his intent to discover and explore lands their own ships had not reached. He therefore concealed his true purpose by explaining Russian explorations in that remote corner of the globe as part of a search for a Northeast Passage.

After several failures to reach the lands he believed lay east of Siberia, Peter planned a more grandiose attempt, even as he felt the onset of what was to become his final illness. "I have been thinking over the matter........[of]finding a passage to China and India through the Arctic Sea," the tsar told the general admiral of his fleet just a few weeks before he died. "I have written out these instructions," he continued, "and, on account of my health, I entrust the execution of them, point by point, to you." Signed on December 23, 1724, a little more than a month before his death, Peter's actual instructions turned out to have nothing to do with a Northeast Passage. Instead, they set in motion the expedition that the Danish sea captain Vitus Bering was to lead overland from St. Petersburg to Okhotsk. From there, Bering was to cross the Sea of Okhotsk to the Kamchatka peninsula, and, after building a ship sturdy enough to navigate the waters of the North Pacific, sail north by northeast to the lands that we now know as Alaska. He was to chart the coast, and collect information that could be used for further Russian enterprises that almost certainly involved colonizing the lands of North America's far northwest.

Aged forty-four when he received Tsar Peter's orders, Vitus Bering was a Dane who had learned his trade on Dutch East India Company ships that had carried him from his native Baltic waters to the Indian Ocean and beyond. Like so many Europeans whom fame had passed by in the West, he had come to Russia to seek his fortune and had received command of his own ship just six years later, in 1710. A brave captain whose modest talent for seamanship won him attention in the navy of a nation whose men preferred to fight on land, Bering seemed much better suited to play the role of explorer than he really was. Stubborn, rigid, and stern, he viewed the world through the prism of an incurious mind, and he had not the passion to explore the unknown that drives true explorers beyond old frontiers. Ready and willing to face adversity, Bering was not a man to shirk his duty-nor did he ever try to reach beyond it. That failing showed in his choice of Martin Spanberg to serve as his second in command. A Dane, like Bering, Spanberg could be counted on to do his utmost to carry out difficult orders, but he had neither vision nor the desire to do more than commanded.

Once described as "a competent officer notorious for his great boorishness and brutality towards subordinates," Spanberg understood Russian poorly, spoke it badly, and barely wrote it at all. A huge dog that he would turn loose upon anyone who made him angry (so rumor had it) went with him everywhere, and like his commander, he interpreted orders in their narrowest sense. Spanberg's dull nature helped to blunt the buoyant daring of Bering's second lieutenant Aleksei Chirikov, who, at twenty-one, was thought to be one of the most promising officers in the Russian navy. He "had a pure heart," one of his superiors wrote, "and a great love of the sea." But Chirikov had graduated from Russia's newly founded naval academy only four years before. He had an explorer's soul, but his inferior rank, limited experience, and tender years obliged him on this, his first voyage, to defer to his gloomy Danish superiors.

Except for their commander, Spanberg, and four others, all of whom caught up three weeks later, Bering's expedition of twenty-six men and twenty-five wagons set out from St. Petersburg on January 24, 1725. Carrying all the cordage and canvas needed to build a ship on the Pacific, they reached the governor's headquarters at Tobolsk in the middle of March and then, after the spring thaws, continued on by river barge to Yeniseisk and Ilimsk, where they spent the winter. The next spring, they portaged across the land bridge that separated Ilimsk from the Lena, built new barges, and made their way to Yakutsk, some twelve hundred miles downriver. There, Bering collected the rest of the supplies and workmen he would need in Okhotsk to cut timber and build a ship. At the beginning of July 1726, he placed Spanberg in command on the thirteen rafts that were to carry the heaviest of their supplies over a network of rivers and portages to Okhotsk. Six weeks later, he set out himself, leaving Chirikov in Yakutsk over the winter with orders to bring the remaining men and supplies overland the next spring.

On January 24, 1725, Chirikov departed with 26 of the 34-strong expedition along the well-traveled roads to Vologda, 411 miles (661 km) to the east. Having waited for the necessary paperwork to be completed, Bering and the remaining members of the expedition followed on 6 February. Bering was supplied with what few maps Peter had managed to commission in the preceding years. Both parties used horse-drawn sledges and made good time over the first legs of the journey. On February 14, they were reunited in Vologda and headed eastwards across the Ural Mountains, arriving in Tobolsk (one of the main stopping points of the journey) on March 16. They had already traveled over 1750 miles. At Tobolsk, Bering took on more men to help the party through the more difficult journey ahead. He asked for 24 more from the garrison, before upping the request to 54 after hearing that the ship the party required at Okhotsk (the

Vostok) would need significant manpower to repair. In the end, the governor could spare only 39, but it still was a significant help. In addition, Bering wanted 60 carpenters and 7 blacksmiths; the governor responded that half of these would have to be taken on later, at Yeniseisk. After some delays preparing equipment and funds, on May 14 the now much enlarged party left Tobolsk, heading along the Irtysh. The journey ahead to the next major stopping point Yakutsk was well worn, but rarely by groups as large as Bering's, who had the additional difficulty of needing to take on more men as the journey progressed. As a result, the party ran behind schedule, reaching Surgut on 30 May and Makovsk in late June before entering Yeniseisk, where the additional men could be taken on; Bering would later claim that "few were suitable". In any case, the party left Yeniseisk on August 12, desperately needing to make up lost time. On September 16, they arrived at Ilimsk, just three days before the river froze over. After the party had completely an eighty-mile trek to Ust-Kut, a town on the Lena where they could spend the winter, Bering traveled on to the town of Irkutsk both to get a sense of the conditions and to seek advice on how best to get their large party across the mountains separating Yakutsk (their next stop) to Okhotsk on the coast.

When Bering reached Okhotsk at the end of October, he found not the seventy-five-year-old town he had been told to expect by his superiors but a squalid settlement of eleven crumbling huts. Without shelter for his men or horses, he set his advance party to work cutting what scant timber could be collected from a land that grew no full-sized trees. Certain that he could not find in the scrubby woods around Okhotsk the timber needed to build an oceangoing ship, Bering decided instead to put his men to work on building a small craft called a *shitik* (the name being derived from the Russian word that means "to sew") that was made by "sewing" planks to a rough-hewn keel with willow twigs and leather thongs. The *shitik* would carry the expedition and its supplies across the Sea of Okhotsk to a new base on the southern tip of the Kamchatka peninsula, where the large forests would yield the first-rate timber needed to build a sturdier vessel that could stand against the storms and waves of the North Pacific.

While Bering struggled to get his advance party settled in Okhotsk, Spanberg got caught in the mountains, his barges frozen in by an early winter storm. Storing the bulk of their supplies as best as they could, Spanberg and his men continued overland, finding shelter at night in dens dug beneath the drifting snow. After losing nearly a dozen of his crew along the way, Spanberg reached Okhotsk at the beginning of January 1727 and spent the next three months sending parties back into the mountains to bring forward the supplies he had left behind. Until Chirikov arrived from Okhotsk with fifty steers and forty more tons of provisions at the end of May 1727, the rest of the men worked at building shelters at Okhotsk and transporting by sled trees that could be cut into ship's timbers from the scrub forests that grew some twenty to thirty miles inland.

By the time Chirikov reached Okhotsk, Bering's shipbuilders had nearly finished the *shitik Fortuna*. After the ship carried its first consignment of shipbuilders and cargo to Kamchatka at the end of June, it returned to Okhotsk in August to take Bering and the rest of his men to Bolsheretsk, a tiny settlement near Kamchatka's southwestern tip that the Russians used as a base for collecting fur tribute. From there, Bering planned to move his entire force across the peninsula to the mouth of the Kamchatka River more than three hundred miles to the northeast. Here, at the rude fort that the tsar's Cossack tribute collectors had built at Nizhnekamchatsk, he planned to lay the keel of the ship that would carry them into the North Pacific the next summer.

To travel from Bolsheretsk to Nizhnekamchatsk required a journey across two ranges of mountains even more dangerous that those that separated Yakutsk from Okhotsk. Several active volcanoes, the highest of which rose to nearly twelve thousand feet and spewed a column of smoke and ashes over a mile into the atmosphere above its summit, made up the easternmost range. The only passage between them was through the Kamchatka River valley, where land turned swampy in summer and temperatures fell to -50 Fahrenheit in the winter. Spanberg led about half of the expedition over that route late in the fall of 1727 so as to begin cutting ship's timber right after the beginning of the next year. The rest followed in two large groups in midwinter, fighting their way through violent Kamchatkan blizzards. "The wind began to blow with great violence, and drifting the snow in quantities, thickened the atmosphere so that we could not see a yard before us," one traveler wrote of a similar storm a century later. "Clouds of sleet rolled like dark smoke over the moor," he added, "and we were all so benumbed with cold that our teeth chattered in our heads." After digging shelters beneath the snow to live through storms such as these, Bering and his men finally struggled into Nizhnekamchatsk in the spring. Now some fifteen hundred miles north of Japan and six thousand miles east of St. Petersburg, they had to build out of green timber a ship strong enough to sail through waters uncharted by anyone in Russia or the West.

The *Vostok* was readied and the *Fortuna* built at a rapid pace, with the first party (48 men commanded by Spanberg and comprising those required to start work on the ships that would have to be built in Kamchatka itself as soon as possible) leaving in June 1727. Chirikov arrived in Okhotsk soon after, bringing further supplies of food. He had had a relatively easy trip, losing no men and only 17 of his 140 horses. On August 22, the remainder of the party sailed for Kamchatka. Had the route been charted, they should have sailed around the peninsula and made port on its eastern coast. Instead, they landed on the west and made a grueling trip from the settlement of Bolsheretsk in the South-West, north to the Upper Kamchatka Post and then east along the Kamchatka River to the Lower Kamchatka Post. This Spanberg's party did before the river froze. Next, a party led by Bering completed this final stint of approximately 580 miles over land without the benefit of the river. Finally, in the spring of 1728, the last party to leave Bolsheretsk, headed by Chirikov, reached the Lower Kamchatka Post. The outpost was six thousand miles from St. Petersburg and the journey itself (the first time "so many [had] gone so far") had taken some three years. The lack of immediate food available to Spanberg's advance party slowed their progress, which hastened dramatically after Bering's and Chirikov's group arrived with provisions. As a consequence, the ship they constructed (named the *Archangel Gabriel*) was ready to be launched on June 9, 1728 from its construction point upriver at Ushka. It was then fully rigged and provisioned by July 9. On July 13, it set sail downstream, anchoring offshore that evening. On July 14, Bering's party began their first exploration, hugging the coast in not a northerly direction (as they had expected) but a north-easterly one.

Sailing further north, Bering entered for the first time the strait that would later bear his name. On August 8, the expedition had a first meeting with the indigenous population. A boat of eight Chukchi men approached the ship and asked the purpose of their visit. They refused to board the ship, but sent a delegate who swam to the ship on an air-filled balloon made from animal skin. The man told that there are islands nearby, and indeed, two days later the expedition reached an island, which Bering named St. Lawrence Island. In turn, Chirikov named the place of meeting the boat as Cape Chukotsky.

After Cape Chukotsky, the land turned westwards, and Bering held a discussion with his lieutenants on August 13, 1728 whether they could reasonably claim it was turning westwards for good: that is to say, whether they had proven that Asia and America were separate land masses. The rapidly advancing ice prompted Bering to make the controversial decision not to deviate from his remit: the ship would sail for a few more days, but then turn back. The expedition was neither at the most easterly point of Asia (as Bering had supposed) nor had it succeed in discovering the Alaskan coast of America, which on a good day would have been visible to the east. As promised, on August 16, Bering turned *Archangel Gabriel* around, heading back towards Kamchatka. On August 31, the ship was hit by a severe storm and barely avoided crashing into the shore. After hasty repairs, on September 2 it reached the mouth of the Kamchatka River, fifty days after it had left.

The mission was at its conclusion, but the party still needed to make it back to St. Petersburg to document the voyage (thus avoiding the fate of Semyon Dezhnev who, unbeknownst to Bering, had made a similar expedition eighty years previously). In the spring of 1729, the *Fortuna*, which had sailed round the Kamchatka Peninsula to bring supplies to the Lower Kamchatka Post, now returned to Bolsheretsk; and shortly after, so did *Archangel Gabriel*. The delay was caused by a four-day journey that Bering made eastwards in search of North America, to no avail. By July 1729 the two vessels were back at Okhotsk, where they were moored alongside the *Vostok*; the party, no longer needing to carry shipbuilding materials made good time on the return journey from Okhotsk, and by February 28, 1730 Bering was back in St. Petersburg. Russia's child Emperor, Peter II, had moved the capital back to Moscow. When Bering addressed Russia's senior officers and the Ruling Senate, one commentator wrote, "some clapped their hands while others shrugged their shoulders." With Peter the Great dead and the secret vision of extending Russia's power to a third continent not yet known to them, the senators delayed voting Bering the thousand-ruble-reward that noted explorers usually received. They let his salary go unpaid for another two years, and when he asked to be promoted to the rank of rear admiral in recognition of his service, the senators declined to act.

However, in December 1731 he was awarded 1000 rubles and promoted to captain-commander, his first noble rank, whereas Spanberg and Chirikov were made captains. It had been a long and expensive expedition, costing 15 men and souring relations between Russia and her native peoples, but it had provided useful insights into the geography of Eastern Siberia, and presented evidence that Asia and North America were separated by sea.

Bering still had a few supporters. A few of them held high posts in the government. These men convinced him to propose an even more ambitious second expedition destined to appeal to the new empress Anna's pretentions for enlarging upon Peter the Great's dream of winning acclaim for Russia among the nations of Europe. This undertaking which history would remember as the Great Northern expedition eventually became so complex that no one man could keep its many strands from becoming tangled. What began as a new attempt by Bering to chart the far northwestern coast of North America soon expanded into what has been described as "one of the most elaborate, thorough, and expensive expeditions ever sent out by any government at any time." As Imperial Russia's first attempt to win acclaim among the geographers and scientists of the West, the Great Northern Expedition was to be as vast in scope as the lands it was sent to explore.

Chapter-13

The Great Northern Expedition

TO SHOW THE WORLD WHAT THE NEWLY EUROPEANIZED RUSSIA OF PETER THE GREAT could accomplish, Empress Anna and her advisers used Bering's proposal as a starting point for conceiving the Great Northern Expedition in the grandest tradition of science. Its nautical program had three main divisions, each of which was a major undertaking in itself, and, which, when taken together, exceeded the scope of any enterprise yet planned by any European nation. Exploring the Kurile Islands on the way, Spanberg was to chart Siberia's eastern seas from Okhotsk to Japan and establish commercial relations with the island kingdom that Europeans had only just begun to enter. At the same time, Bering was to sail across the North Pacific from Kamchatka to find the northwestern coast of North America. He was to chart it as far north as possible, and make contact with the natives with an eye to collecting fur tribute and finding new supplies of food for Siberia's far eastern lands. Then, while these enterprises were being launched from Siberia's Pacific coast, half a dozen other expeditions were to chart seven thousand miles of Siberia's shoreline, working their way east from Arkhangelsk on Russia's White Sea coast until they reached Bering Strait, where they were to turn south and follow the coast of the Chukhotka peninsula to Kamchatka.

Blocked by massive ice fields for much of the year, most of the coastline to be charted by the Great Northern Expedition lay above the Arctic Circle. The men and ships of the Hudson's Bay Company would spend the better part of two centuries on a similar mission in North America, and they would not finish their work until 1847. But, with a degree of daring that can come only from being willing to sacrifice men's lives to win greater glory, the organizers of the Great Northern expedition planned to finish all of their work in a single decade. Nor did they intend to stop at exploring Siberia's seas. As part of this immense undertaking, a contingent of learned men attached to the Great Northern Expedition from the Imperial

Russian Academy of Sciences was to produce nothing less than a complete historical, physical, botanical, ethnographical, and linguistic description of Siberia. For a time, some of the men in the Ruling Senate talked of having part of the expedition explore Central Asia, and the Admiralty even considered adding a voyage around the southern tip of Africa to the enterprises Bering could command.

As set down by the Ruling Senate, the Admiralty and the Academy of Sciences, the tasks of the Great Northern Expedition were more visionary than precise. They were entirely without clear organizing principles or a workable structure of command, and so vast in scope that no single hand could control them all. Like a roomful of mad political alchemists, the expedition's planners mixed the principles of democracy and tyranny in the decision-making process of each enterprise. They made the officers sent to chart Siberia's Arctic coasts directly responsible to the Admiralty for the success of the enterprises they commanded but allowed their crews to make vital decisions by majority vote. At the same time, the expedition's "learned guard", from whom the Ruling Senate instructed Bering to seek advice in all things, were to answer only to their colleagues in the Academy of Sciences. Commanders of the expedition's separate enterprises might turn to Bering for more men, supplies and equipment but control of the funds that purchased them remained in the hands of the Ruling Senate in St. Petersburg. Only the rigid stance with which the Admiralty faced every obstacle kept the diverging strands of the Great Northern Expedition from coming apart. Sickness, accidents, faulty equipment, and unexpected tragedies all counted for nothing in the minds of the men who directed the Great Northern Expedition from the chanceries of faraway St. Petersburg. No matter what obstacles the explorers exported, the lords of the Admiralty always gave the same reply. Somehow, in some way, "the work must be done."

Just as the Russian's cost of Siberia's conquest had been taken from Siberia itself, so the statesman of St. Petersburg expected to pay the expense of Siberia's exploration from the same source. Originally estimated to require no more than twelve thousand rubles from the Imperial Treasury because most of the supplies were to come from government storehouses along its route, the Great Northern expedition had cost the Russians more than three hundred thousand rubles before its work was halfway done. Then all the doubts that had plagued Russia's senior statesmen about their dour Danish admiral after his first expedition came surging back. Could something not be done so that "the Treasury should not be emptied in vain?" the empress's cabinet asked at one point. Would it not be possible, they petitioned the Admiralty, "to look into the Kamchatka [that is, the Great Northern] Expedition and see if it can be brought to a head?" Now embroiled in expensive wars with the Ottoman Empire and Poland, Russia's senior statesman had lost some of their peacetime enthusiasm for science and exploration and worried instead that the Great Northern Expedition might strain the resources of the Imperial Treasury too far.

The costs of the Great Northern Expedition had soared because so many parts faced such huge difficulties in completing their assignments, especially in charting Siberia's Arctic coast. Unpredictable, treacherous, and controlled only the whims of wind and temperature, the ice at those latitudes never disappeared. It only broke apart enough between the middle of June and the end of September for water to appear in the spaces that separated one iceberg from another. Russia's explorers had to work their fragile wooden ships through these perpetually shifting mountains of ice and, while doing so, take all the readings needed for charting the coast along which they sailed. At most, they had four months out of twelve in which to work-four months during which sunlight at midnight tempted them to go without sleep in order to extend the working time that the return of winter would cut short. Then the great frozen masses that had floated free for the summer became fields of solid ice that closed all navigation. Frozen fast in the ice, the small oar and sail-driven ships of the Great Northern Expedition's explorers would then have to wait for the next summer to free them to continue their work.

On the advice of the Ruling Senate, the Admiralty, and the Academy of Sciences, the Great Northern Expedition's planners divided Siberia's coastline into four segments, starting at Arkhangelsk in the west and working east. The coast from Arkhangelsk to the mouth of the Ob comprised the first segment, from the Ob to the Yenisei the second, from the Yenisei to the Lena the thirds, and from the Lena to the mouth of the Kamchatka River on Siberia's Pacific coast the fourth. Each had its own special perils, almost none of which the Russians foresaw. Between Arkhangelsk and the Ob, the icebergs of the fog-filled Kara Sea created a gantlet that even the most determined explorers found difficult to overcome. The same obstacles filled the passage from the Ob to the Yenisei. East of the Yenisei, the risk of being crushed by ice increased, for the explorers had to sail even further north to round the Taimyr peninsula, the northernmost point of Eurasia. Then, from the Taimyr peninsula to the delta of the Lena River, the ice got worse, for the colder winters to the peninsula's east meant that the coastal waters of the Laptev Sea remained frozen well into July. From the Lena to Bering Strait, the desolation of Siberia's northeastern coast heightened the danger of fog-blanketed seas filled with shifting mountains of ice. Although some of the first Russian *promyshlenniki* and traders had navigated these waters in their tiny flat-bottomed boats during the first half of the seventeenth century, they had left no charts.

The westernmost enterprise of the Great Northern Expedition was to chart the coast of northeastern Europe and western Siberia from the White Sea port of Arkhangelsk eastward to the Ob River town of Berezov. Berezov was once a center of the Arctic fur trade but now shrunk into obscurity. This should have meant little more than following the route of the small coastal traders who had sailed without compasses, astrolabes, or sextants between Arkhangelsk and Mangazeya at the beginning of the seventeenth century. The task proved more difficult than expected because the Kara Sea had greater quantities of ice in the days of Bering than when Mangazeya and Berezov had flourished. When naval lieutenants Stepan Muravev and Mikhail Pavlov set sail from Arkhangelsk in the summer of 1734 on a voyage that had taken from four to six weeks in the days of Siberia's first fur rush, they therefore got no farther east than the mouth of the Pechora River by fall. Less than halfway to their destination by the time the sea began to freeze, they spent the winter in Pustozersk, a lonely outpost to which a number of famous Russians had been condemned to exile in years gone by. The next summer, Muravev and Pavlov sailed a short distance east, but masses of icebergs that lay shrouded behind fogs of the Kara Sea forced them back to Pustozersk for another winter.

Like many men of equal rank, Muravev and Pavlov were fierce rivals, and, by the time they returned to Pustozersk in the fall of 1735, their rivalry had turned to hatred. Charging each other with incompetence, corruption and brutality, they flooded the Admiralty with complaints only to have their superiors reduce them both to the rank of common seaman "for numerous dishonorable, slothful and stupid actions." The work must be done and progress reported, the Admiralty insisted. There was no room in his plan for personal squabbles that slowed the pace of work that was to win acclaim for Russia in the West.

In the spring of 1736, the Admiralty sent Stepan Malygin to replace Muravev and Pavlov. Before early frosts forced him into winter quarters in the middle of September, Malygin sailed his ships into the Kara Sea. He reached the Southwest coast of the Iamal peninsula. From there he continued to Berezov the next summer. Based on the detailed reading he had taken, Malygin could report that the southern tip of Novaia Zemlia, the landmass that rose from the sea north of the Urals, was neither a peninsula nor another continent but an island and that the Iamal peninsula itself was not an isthmus leading to land further north. Hundreds of newly discovered islands still had to be explored when Malygin made his report to the Admiralty, but the first segment of the Great Northern Expedition's work was done by the summer of 1737. If the ships and crews working on the remaining segments of Siberia's Arctic coast enjoyed similar success, the navigators and cartographers of the world would come to know its details long before they learned the particulars of the northern coast of North America.

Information was beginning to come in from segments of the expedition that were working farther east, but it would take more time to finish their work and evaluate its success. Sailing down the Ob River from Tobolsk with a crew of fifty-six in the summer of 1734, the young naval lieutenant Dmitri Ovstyn set out in the *Ob Postman*, an oddly named seventy-foot ship from the deck of which he hoped to chart the few hundred miles of Arctic coast between the Ob and the Yenisei in a single summer. Ice in the Kara Sea drove him back almost as soon as he left the Ob. After spending the winter in Obdorsk, a small settlement at the river's mouth, he tried again in the summer of 1735, but scurvy struck his crew with a vengeance and forced him to wait again. After doing so better in the summer of 1736, Ovstyn finally reached the Yenisei and sailed upriver to Turukhansk in 1737. Then, after taking his ship and crew onto Yeniseisk the next spring, he hurried to St. Petersburg to make his report. The second segment of charting Siberia's Arctic coast was done, but very bad news awaited Ovstyn in the Russian capital.

Between 1725 and 1741, St. Petersburg became a morass of political intrigue that brought seven sovereigns to the throne in sixteen years. Unaware of how the currents of Russian politics had shifted during his four-year absence, Ovstyn became the victim of one of the many upheavals that caused this time in Russia's history to be known as the "Era of Palace Revolutions". Having been a friend of Prince Aleksei Dolgorukov, who had fallen from favor during his absence, Ovstyn was, like Muravev and Pavlov, reduced to the rank of common seaman and sent to Okhotsk to join the expedition that Bering and Chirikov were preparing to lead to Alaska. There, he would become Bering's aide, a sailor whose experience in Arctic waters far exceeded that of his chief. When he returned, he would fall into obscurity.

Although the work of the expeditions charting Siberia's coast from Arkhangelsk to the Yenisei had been slowed by weather and misfortune, greater obstacles faced the men assigned to chart Siberia's Arctic coast east of the Yenisei's mouth, where the Taimyr peninsula stretched far out into the icy waters of the Arctic to divide the Kara and Laptev seas. Almost perpetually icebound, this coast had to be charted if the Great Northern Expedition was to complete its work, but, as far as navigators knew at that time, no ship had ever rounded its northernmost cape from either east or west. Deciding to approach the peninsula from both directions at once, the Admiralty assigned Fedor Minin to sail around it from the west at the same time as they ordered Vasili Pronchishchev and his redoubtable wife, Maria, the world's first female polar explorer to sail toward it from the east.

At the beginning of June 1738, Minin and a crew of fifty-seven boarded the *Ob Postman*, which Ovstyn had just safely brought into Yeniseisk. After following the Taimyr peninsula's coast to the northeast and passing its northernmost point, Minin's optimistic instructions from St. Petersburg read, he and his crew were to turn south and sail on to the mouth of the Khatanga River. If there were time, they should return to the Yenisei by the same route. Otherwise, he and his men should winter on the Khatanga and return the following summer. For most of July, Minin's expedition fought ice in the long bay that separated the Yenisei's mouth from the Kara Sea until heavy fog forced them to stop. By the time a south wind cleared the fog and allowed them to sail on, the water in the ship's casks were beginning to freeze and snow was starting to fall. When a boat sent ashore reported that the streams and small rivers had already frozen, Minin knew he would have to put his crew ashore to build winter quarters without ever having reached the Taimyr peninsula.

The next summer brought no more progress. After storms and ice drove his ship back into the mouth of the Yenisei, Minin sent his second in command to chart the Taimyr coast by land. He returned empty-handed after the cold had frozen his instruments and the blazing light reflected from the Arctic snow had injured his eyes. In the summer of 1740, Minin and his crew fought their way through the icebergs to the mouth of the Piasina River, perhaps a hundred and fifty miles north of the winter quarters they had built two years before. By that time, his superiors had decided to replace him, despite the maps and charts he had been sending to St. Petersburg of the small segments of Siberia's coastline he had managed to explore. After the lords of the Admiralty relieved him in the fall of 1740, they debated his case for nearly a decade before they reduced him (like Ovstyn, Muravev and Pavlov) to the rank of common seaman. Disgrace and punishment were becoming the regular rewards of even the most daring men who failed to conquer the ice and weather in Siberia's Arctic waters.

In the meantime, naval lieutenant Vasili Pronchishchev, his wife, Maria, and a crew of fifty-five were trying to round the Taimyr peninsula from the east in order to reach the ice-filled seas that had defied the best efforts of Minin's crew. Sailing in a near carbon copy of the *Ob Postman* called the *Iakutsk*, they had left Yakutsk at the end of June 1735 and sailed down the Lena River to the sea. They had then begun to work their way west. Two months and hundred miles later, they reached the mouth of the Olenek River. There they built winter quarters and began to transfer the sightings and soundings they had taken during the summer onto detailed charts. The next summer, ice kept the *Iakutsk* bottled up in Olenek harbor until the beginning of August. By the time the ship reached the Khatanga River some three hundred miles west, both of the Pronchishchevs were dying of scurvy. The expeditions remaining officers and crew then voted to retreat to Olenek for a second winter because "there were neither people nor any wood" to be found on the Khatanga. By the time they reached winter quarters, both of the Pronchishchevs were dead. The next summer, their chief navigator semen Cheliuskin took the remnants of the expedition back to Yakutsk. The Admiralty then transferred command of the *Iakutsk* to Khariton Laptev, an officer who had just been promoted to junior lieutenant.

We have no record of when or where Laptev was born or what sort of family he came from. He was an immensely resourceful man who had been blessed with a deep understanding of what moves men in times of stress. Laptev knew that only proper rations, enough equipment, and a crew dedicated to their work could bring success in the Arctic. He therefore asked the Admiralty to pay his men before they sailed and build storehouses for extra supplies along the course he expected to chart. The Admiralty agreed to all his requests, but no amount of material help could modify the terrible climate that Laptev and his men had to face. Only in July and August (according to twentieth century calculations) were the average temperatures in the Taimyr peninsula's northern lands above freezing. For the rest of the year, they ranged from -21 to-31 Centigrade (November through April) to a high of -17 Centigrade in June.

Like the Pronchishchevs, Laptev had to fight ice every foot of the way in the sea that now bears his name. Inching between the shifting floes and icebergs, the *Iakutsk* worked her way back to the Pronchishchevs winter quarters on Khatanga Bay but then, after sailing north for just eight days, had to turn back. Finding neither food nor driftwood on shore, Laptev decided to winter back on the Khatanga, where men sent overland from Yakutsk had already built winter shelters and laid in supplies for his men. The next summer, while Minin was making his last effort to work his way up the Taimyr peninsula's western coast, Laptev tried again. This time, icebergs crushed the *Iakutsk* and drove him and his crew back to their winter quarters at Khatanga Bay on foot.

With the *Iakutsk* gone and no means of replacing her, Laptev ordered Cheliuskin (the same pilot who had guided the Pronchishchevs) to take a small party across the peninsula on dogsleds and proceed north from the point of Minin's farthest advance. Five weeks later, at the end of April 1741, Laptev himself set out with a single soldier and a Yakut guide to meet Cheliuskin, and the two spent the summer charting the peninsula's western coast by land. After retreating nearly five hundred miles to the shelter of Turukhansk for the winter, Cheliuskin set out again and finally reached the northernmost tip of the Taimyr peninsula at the beginning of May 1742. After that, he rejoined Laptev and the two men returned to St. Petersburg to make their report to the Admiralty. Now that Eurasia's Arctic coast had been charted from Arkhangelsk to the mouth of the Lena River, both men agreed that even if there turned out to be no land barriers farther east, the hitherto impenetrable ice around the Taimyr peninsula made Siberia's Northeast Passage impassable.

While Laptev and Cheliuskin had been struggling to chart the coast of the Taimyr peninsula, another of the Great Northern Expedition's crews had been trying to make headway farther east. When the Pronchishchevs had sailed the *Iakutsk* north down the Lena to begin their explorations in June 1735, a slightly smaller sister ship, the *Irkutsk,* had followed in their wake under the command of naval lieutenant Petr Lasinius. Lasinius had then sailed the *Irkutsk* east to follow Dezhnev's still unknown course to the Bering Strait and the Anadyr River, but he had none of Dezhnev's good fortune. Icebergs kept him bottled up in the Lena delta until the beginning of August and then stopped him just four days after he finally worked his way into the Laptev Sea. Scurvy killed Lasinius in December and left fewer than twenty of his crew alive by spring, when Dmitri Laptev, a cousin of the young lieutenant who had taken the Pronchishchevs place on *Iakutsk,* arrived to take command.

Like Lasinius, Laptev made no headway at first. Convinced that the task was hopeless, he went back to St. Petersburg in the spring of 1737. Here his superiors in the Admiralty told him that the work must still be done, if not in one year, then in two or three, and that, if the coast could not be charted by sea, then it must be done by land. Laptev returned to Siberia, took the *Irkutsk* to sea once again, and by the summer of 1739 had advanced somewhat more than four hundred miles to the mouth of the Indigirka River before the ice closed him in. The next year, Laptev fought his way through the ice to the Kolyma River and spent the winter building smaller boats that he hoped might be more maneuverable in the ice-filled seas farther east. When these managed to advance barely twenty-five miles in the summer of 1741, Laptev decided to go overland to Anadyrsk, the trading center that Dezhnev had founded on the southern coast of the Chukhotka peninsula after his shipwreck almost a century before. From Anadyrsk, Laptev tried again to approach Siberia's easternmost tip in 1741. But, although he and a number of others managed to reach its shores, its waters could not be charted until 1823, when Lieutenant Ferdinand Vogel and a group of surveyors working from dogsleds finally mapped the last fragments of Siberia's northeastern coastline.

Except for the work done by Vrangel seventy years later, the contingents of the Great Northern Expedition that the Admiralty had assigned to chart Siberia's Arctic coast proved that the Northeast Passage, which men had dreamed about since the sixteenth century, did in fact lie along Siberia's northern coast. The obstacle to sailing that route was ice, not land, and, for the time being, that remained as formidable a barrier as any land could be. But the day would come at the end of the nineteenth century when Siberia's Northeast Passage would be opened to ships of the East and West. In 1875, for the first time, Sweden's Baron Nils Adolf Nordenskjold sailed in the steam-powered Vega from Stockholm through Bering Strait, and by the early twentieth century timber from the Lena was being shipped regularly to mills in England by way of the Northeast Passage. In Soviet times, use of this route would become too common for comment. Perhaps most amazing of all, when Nordenskiöld took readings with his more precise modern instruments in 1875, the difference between them and those that Cheliuskin and Laptev had taken from their dogsleds proved to be less than three minutes. In that sense, Russian science had accomplished more during the 1730s and 1740s than anyone had dared to hope. But the Great Northern Expedition's achievements did not end there. While men in fragile wooden ships had been fighting to make headway against the fog, ice, and cold of Siberia's northern coast, others had been at work in its interior in the largest scholarly venture to be undertaken by any group of scientists before the nineteenth century.

In 1733, the newly born Imperial Russian Academy of sciences had contributed a detachment of scholars to the Great Northern Expedition and assigned them to produce a complete historical, physical, botanical, ethnographical, and linguistic description of Siberia. Four men, three of whom were not Russian, formed the core of this enterprise, and at their head stood Gerhardt -Friedrich Muller. An introspective Westphalian scholar who, although only twenty-eight, was known to suffer prolonged fits of morbid hypochondria, Muller went to Siberia to describe the customs and languages of all of the native tribes living between the Urals and Kamchatka and to search through whatever archives could be found to write the history of the Russian conquest. While he did so, Johann Georg Gmelin, a brilliant twenty-two-year-old Wurttemberg physician, professor of chemistry, and passionate student of natural history, assembled materials on Siberia's flora and fauna. Less accomplished and somewhat older, their third companion was a mediocre French scientist by the name of Louise Delisle de la Croyere, who, despite a knowledge of astronomy that has been described as "very defective", had been assigned to determine the precise latitude and longitude of Siberia's main towns and landmarks. All three men looked down upon their Russian assistant, a young naturalist by the name of Stepan Petrovich Krasheninnikov. Two years younger than Gmelin and one of the first Russians to study at the Academy of Sciences in St. Petersburg, Krasheninnikov would return after eight years in the Siberian wilderness to become as renowned as any of the Europeans who had disdained him.

Supported by an impressive host of artists, topographers, and scribes and traveling with a jumble of scientific paraphernalia including telescopes up to fifteen feet in length that de la Croyere insisted he could not work without, the academicians set out for Siberia in August 1733. It took them six months to reach Tobolsk and another six to reach Yeniseisk. Now that the Russians had become firmly settled in Siberia, travel no longer had to be limited to the spring, summer and fall. The academicians therefore moved on from Yeniseisk to reach Irkutsk at the beginning of March 1735. By then, Muller and Gmelin were squabbling with the local authorities about the quality of their accommodations and were finding east Siberian life distinctly less than pleasant. They spent most of the summer in Kyakhta, studying how this newly built gateway to China affected Russia's trade, and then returned to Irkutsk for the winter. Only in September 1736 did they finally reach Yakutsk, which was supposed to become the base from which they were to explore the peninsula of Kamchatka, the last part of Siberia to be occupied by the Russians.

None too pleased with the primitive conditions under which they had been obliged to live and work in Irkutsk and Yakutsk, Gmelin and Muller (who carried bottles and casks of their favorite European wines with them wherever they went) decided to set aside the part of their assignment that required them to go on to Kamchatka, where Russia's grip was the weakest and the conditions the least comfortable. They concentrated on their studies of history, ethnography, and botany farther west. Muller spent the next decade making forays to the archives of remote Siberian chancelleries and monasteries from headquarters in the more comfortable, congenial atmosphere of Yeniseisk and Tobolsk. Gmelin collected the specimens and made the observations needed to prepare the text of his massive four-volume *Flora Sibirica*, which is still remembered as a classic in the history of botany. When the two men returned to St. Petersburg toward midcentury, they had travelled some twenty-three thousand miles. Muller had copied thousands of rare documents that would make it possible for him to tell Siberia's story as part of the Russian's first effort to describe their national past.

Everywhere he travelled, Muller unearthed historical treasures that showed how the Russians had taken Siberia and at what cost. At

Solvychegodsk, he found documents in the Stroganov's private archives that told the details of Yermak's campaign. Among the long-forgotten documents in the government storerooms in Yakutsk, he unearthed Dezhnev's amazing report describing how he and a handful of Cossacks had done almost a century earlier what Lasinius and Laptev now had failed to accomplish. After verifying its age and authenticity, Muller immediately pronounced Dezhnev's aging file to be genuine, but others, citing the difficulties that Lasinius and Laptev had just encountered, insisted that such a feat was impossible. Skepticism about the Dezhnev file grew as the years passed, especially as explorers with larger ships, more training, and better equipment continued to fail in their efforts to round Siberia's East Cape. Not until well into this century, when the best Russian experts finally proved that Dezhnev and his handful of Cossacks had done precisely what they claimed to have done, could the debate be put to rest. In the meantime, Muller outlived seven Russian autocrats before he died at the age of seventy-eight in 1783. Using the notes and materials about history, geography, languages, and ethnography of the lands and peoples east of the Urals that he and Gmelin had assembled, Muller, the German who spent nearly all of his adult life in Russia, became the father of Siberian history.

While Muller and Gmelin studied the history of Siberia, the life and languages of its natives, and the flora and fauna between the Urals and Yakutsk, they had sent their assistant Krasheninnikov on to Kamchatka so as to avoid going there themselves. Armed with detailed instructions about what he was to investigate and how, Krasheninnikov first travelled from Yakutsk to Okhotsk in nothing resembling the grand style of his superiors, fighting the terrain and climate as he went. Nearly a hundred years had passed since the first Russians had made their way across the mountains to found Okhotsk. The passing of time had not made the journey appreciably more pleasant or less dangerous. As he and the handful of men who accompanied him clambered across boulder-filled streams and worked their way across bog-stopped mountains, Krasheninnikov felt himself almost entirely at the mercy of the elements. "On the summits [of the mountains], there are terrible bogs and quagmires," he later wrote. "One is terrified," he added, "by the way the earth there quivers." Everywhere, the land seemed determined to stand against any man who tried to cross it. It would be very hard, Krasheninnikov remembered many years later, "to imagine a more difficult crossing."

Things got no easier once Krasheninnikov reached Okhotsk where the ancient *Fortuna* that Bering had used to ferry his first expedition to Kamchatka was still in service. Now much the worse for wear after a decade of carrying cargo between Okhotsk and Bolsheretsk, the *Fortuna* began to take on water soon after she left port with Krasheninnikov and his supplies and instruments abroad. Less than twelve hours after the ship had left Okhotsk, the men working the pumps in her hold were standing in water up to their knees, and all the crew were bailing with kettles, and anything else they could find. Only by throwing the entire seven tons of cargo overboard could the crew lighten the *Fortuna* enough to get her to Bolsheretsk. Here, a storm stirred by an earthquake immediately drove her against the shore. "The next day, we found nothing but planks from the wreckage of our ship," Krasheninnikov reported. "Inside," he added, "they were all so black and rotten that one could break them by hand without difficulty." Krasheninnikov now had no supplies, almost no instruments, and his only wits to support him, since Bering insisted that he had few supplies to spare.

Arriving at Bolsherestsk late in October 1737, Krasheninnikov worked on his own for much of the next four winters and three summers. Wandering through the wilds of Kamchatka, he collected the specimens needed to piece together a full-scale portrait of the primeval land of earthquakes, volcanoes, geysers, quaking bogs, tidal waves and avalanches into which his passion for science had brought him. He charted Kamchatka's rivers, mapped its trails and passes, crisscrossed its many mountains, discovered its hot springs, and struggled to understand the materials he had gathered and the things he had seen. "It has neither grain nor livestock and suffers frequent earthquakes, storms and floods," Krasheninnikov wrote of Kamchatka in the voluminous account he published some fifteen years later. "Almost the only diversion," he added, "is to look upon its towering mountains, whose peaks are continually covered with snow, or, if one lives along the sea, to listen to the crashing of waves and observe the ways in which different species of sea animals live in hostility and friendship with each other." This was hardly an enthusiastic portrait of the land in which Krasheninnikov spent more than three years, but from the point of view of ethnography and natural history, Kamchatka was a veritable wonderland, and it was upon that that the young scientist concentrated.

What fascinated Krasheninnikov most about Kamchatka were the Kamchadals. The Kamchadals lived on the southern three quarters of the peninsula. During the centuries of isolation, they had managed to shape innocence and perversity into a way of life. "The natives of Kamchatka are as wild as the land itself," he wrote. "They are filthy and vile [and]...they are so infested with lice...that, using their fingers like a comb, they lift their braids, sweep the lice together into their fists and then gobble them up." As one of Russia's recently Europeanized elite and a man who enjoyed the good food, comfort, and leisure that life in the civilized world offered, Krasheninnikov found it hard to endure the society of the Kamchadals. He watched in disgust as they stuffed strips of seal blubber into the mouths of guests and insisted that they be swallowed whole, and he gagged upon delicacies made from fish that had been allowed to rot until the flesh had become gelatinous. "The stench," of such concoctions, he wrote was sometimes "almost impossible to tolerate", yet he did so in order to better understand the people he studied and learn what impact their diet had upon the human organism.

Appalled by the Kamchadal's filth and unable to watch them prepare some of their meals "without feeling sick", Krasheninnikov still found virtue in their savage way of life. "They know nothing of wealth, honor, or glory," he reported, "[and]they do not suffer from greed, ambition, or pride. All of their desires," he continued, "are directed toward enjoying an abundance of whatever they want." The Kamchadals's measure of happiness therefore differed from that of civilized folk. "They think that it is better to die than not to be able to live in the manner they choose," he concluded. The Kamchadals used suicide, Krasheninnikov explained, as "a final measure to achieve happiness."

Krasheninnikov continued his lonely scientific vigil on Kamchatka until the fall of 1740, when Georg Wilhelm Steller, a physician trained at the University of Halle and a brilliant student of natural history arrived from Central Europe. Steller had been a late addition to the contingent that the Academy of Sciences had contributed to the Great Northern Expedition, having reached St. Petersburg from Halle (with a number of less than respectable stops in between) only in 1735. Winning favor with some of the Academy's patrons almost immediately, Steller had set out for Siberia at the beginning of 1738. On the way to Okhotsk, he had met with Muller and Gmelin in Yeniseisk and had been given their blessings for going on to Kamchatka and joining Bering. As part of the bargain, Steller had agreed to oversee Krasheninnikov work. He had then proceeded slowly, collecting specimens and making meticulous notes, until he reached Yakutsk in the spring of 1740. Finding that town unpleasant and the authorities unhelpful, he was anxious to leave almost from the moment he arrived. "In the wickedness, incredible deceit, perfidy and hypocrisy," he wrote of the people he met in Yakutsk, "they are as far removed from the other Siberian inhabitants and the native Russians as are the serpents from the doves." Steller therefore hurried over the mountains and reached Okhotsk in the middle of August.

Therefore, he had his first meeting with Bering and impressed him immediately. Abrasive, dogmatic, usually without tact, and obsessed by his desire to explore the Alaskan coast, Steller could also be charming when the occasion demanded. He soon won Bering's confidence and made certain of a place on one of the ships he would sail to North America, but the problems of Krasheninnikov still remained. Steller had no intention of sharing his newly acquired place as Bering's chief naturalist, least of all with someone whose long study of Kamchatka and its natives might well have made him able to challenge the German's knowledge about the flora and fauna in that part of the world.

Steller therefore saddled Krasheninnikov with all sorts of burdensome tasks and on at least one occasion insisted that he set his work aside and travel the length of Kamchatka just to translate into Russians a number of suggestions about governing Kamchatka and Christianizing its natives that he wanted to submit to the Ruling

Senate. In this and other ways, Steller could make his Russian rival's life a torment. What he could not do was order him back to the mainland. For that, he enlisted the aid of de la Croyere, the mediocre French astronomer and geographer who had parted company with Muller and Gmelin at Yakutsk in the fall of 1736 and had gone on to join Bering's expedition at Okhotsk. De la Croyere's high rank as a professor in the Academy of Sciences gave him the authority to dictate Krasheninnikov's fate. He was more than ready to issue the order Steller wanted.

Evidently not relishing the prospect of working as Steller's humble assistant and, in any case, probably not reluctant to leave Siberia after eight lonely years, Krasheninnikov made no complaint. His departure left Steller and de la Croyere alone to sail with Bering's expedition across the North Pacific in the summer of 1741. But there was still a final irony that would only become evident some years later. Although Steller would have the honor of being the first man of science to set foot on Alaskan soil, it would be Krasheninnikov who, when Steller died from acute alcoholism at the age of thirty-seven, fleshed out the gaps in his notes and published them as part of his own huge study about Kamchatka. Krasheninnikov therefore had a key part to play in reporting some for the scientific findings of the enterprise that history remembers as Bering's second voyage, or the Russian discovery of America.

Chapter-14

The Russian Discovery of America

EVER SINCE YERMAK HAD CROSSED THE URALS, the Russians had moved east, claiming new subjects for the tsar and scouring the Siberian countryside for furs. Then, as the tsar's Cossacks worked their way into the Kamchatka peninsula and struggled to put down roots in the infertile soil around Okhotsk, only water stood on the horizon. What lay beyond Siberia? How far across the Pacific was it to North America from Kamchatka? What lay in between? Were there new lands from which fur tribute could be collected as the returns on the Siberian mainland continued their precipitous decline? The prospect of planting Russia's flag on a third continent was dazzling enough to induce Peter the Great, newly crowned head of the youngest of Europe's empires, to assign precious men and ships to exploring the North Pacific the moment the pressures of the Great Northern War with Sweden eased. Two expeditions had preceded Bering's first voyage. Like his, neither had reached the New World.

Just three years after Bering had turned back from the voyage that had carried him through Bering Strait to within three day's sail of the Alaskan coast, the pilot Ivan Fedorov and the surveyor Mikhail Gvozdev set sail in the *St. Gabriel* in search of a "big country" of great forests and high mountains that, Siberia's natives said, lay scarcely more than a day's voyage beyond the ocean. Less than a month after they sailed from Kamchatka on July 23, the *St. Gabriel*'s crew reached the coast of the "big country" but could not land because of the head winds and shallow water. "We could see huts," Gvozdev remembered, "but in spite of our best efforts we did not come as close to them as we wished." Then, "because of the lateness of the season and the stormy weather," Fedorov and Gvozdev sailed back to Kamchatka, still thinking that the "big country" was an island and not knowing that they had discovered

America from the west.

A few months later, Fedorov died. Soon after that, the ship's log disappeared, and the brief report that Gvozdev claimed to have sent to the authorities at Okhotsk seems never to have been forwarded to St. Petersburg. As the architects of the Great Northern Expedition made plans for Bering and Chirikov to search for North America once again, none of them knew that the "big country" and the northwestern corner of North America were one and the same, or that the *St. Gabriel* had reached it six years earlier. Still thinking that the "big country" was an island, the lords of Russia's Admiralty assigned Bering the task of sailing beyond it to North America.

Having decided to carry out their mission as soon as ships could be built at Okhotsk, Bering and Chirikov set out from St. Petersburg in the spring of 1733. Officially, in addition to the expedition he was to lead to the coast of North America, Bering also had responsibility for Spanberg's explorations of the Kuriles and the waters north of Japan. He was also accountable for every ship that sailed to chart Siberia's Arctic coast, for reforming the administration of Siberia, and for all of the work that the contingent sent by the Academy of Sciences was to accomplish. Only a man of stupendous energy, supremely Machiavellian outlook, and immense scientific genius could even have dreamed of controlling such an enterprise. Bering was by no stretch of the imagination such a man. Nor would the plan according to which his superiors had organized all these undertakings have permitted him to exercise any real authority except over the voyage across the North Pacific that he himself was to command. But even though they limited Bering's control over the Great Northern Expedition, its planners made certain that he was well positioned to take the blame for its failings. As the work slowed during the next half decade and the costs soared, the Ruling Senate hurried to accuse him of embezzlement, dealing in contraband liquor, and any number of other crimes. Not one of its charges ever proved to have any substance. In the meantime, the senators withheld Bering's pay for the better part of two years and continued to threaten him with court-martial.

While he struggled to defend himself against these charges, Bering faced problems with the preparations for the expeditions that he, Chirikov and Spanberg planned to lead into the Pacific. While he and Chirikov had been assembling a full complement of men and supplies at Yakutsk during the summer of 1735, Spanberg had gone ahead to Okhotsk only to find that the barracks promised for the expedition had not yet been started, that hardly any supplies had been laid in, and that the timber that was supposed to have been ready to build their ships had not even been cut. Expecting that the ships that he, Bering, and Chirikov needed would be launched by 1736 or 1737 at the latest, Spanberg began to build barracks at Okhotsk instead.

In disgust, Bering hurried to join him. "We have sand and pebbles, no vegetation whatever, and no timber in the vicinity," he wrote after he had surveyed the situation. "Firewood must be obtained at a distance of four to five miles, drinking water one to two miles, while timber and joints for shipbuilding must be floated down the river twenty-five miles. It was easy enough to fix the blame on officials whom the admiralty had sent to prepare the way, and Bering certainly did his share of finger pointing before his ships were ready to sail. But he was dealing with a navy that was less than half a century old and a country in which the tradition of corruption and shifting blame was as old as time itself. In the meantime, shipbuilders had to be put to work on cutting timber and building the barracks that should already have been finished. Only then could they move on to laying the keels for their ships.

To make matters more difficult for the crews and officers who were to sail to North America, the first resources went to Spanberg's expedition because it had been scheduled to sail before Bering. With two new ships to sail along with the old *St. Gabriel,* Spanberg finally set out for the Kuriles in June 1738 and, the next year, sailed farther south to prove that the Kuriles formed a chain that stretched from Kamchatka to Japan. Yet Spanberg proved to be so careless in keeping his ship's logs that the lords of Russia's Admiralty had to order him to repeat his explorations because they found it "quite impossible" to chart the course he had sailed from the readings he had taken. Then, further misunderstandings with his superiors brought Spanberg to the brink of death three years later. Had the

Danish government and several influential friends not interceded, he would have ended his career on the gallows.

While Spanberg was exploring the Kuriles, help arrived for Bering in the form of two adjutants sent by the empress with full authority to impose his requests for men and supplies upon the uncooperative Siberian officials who had been standing in his way. By the end of June 1740, a pair of two-masted ships, each measuring eighty feet in length and carrying fourteen small cannon had been launched at Okhotsk. That October, Bering and Chirikov sailed to Petropavlovsk, the harbor that was being built on eastern Kamchatka's Avacha Bay. Once the supplies that had been landed on the peninsula's more sheltered Okhotsk sea coast were carried across the mountains during the winter of 1740-41, the second Bering expedition was ready to sail.

When the *St. Peter* and *St. Paul* sailed out of Avacha Bay on June 4, 1741, each ship carried a crew of seventy-six, a hundred casks of water, about sixty-five tons of beef, pork, groats, flour, butter and salt, and a list of signals to help each to keep in touch with the other. Chirikov commanded the *St. Paul*, to which de la Croyere had been assigned, while Steller, his arrogantly eccentric personality already having put him at loggerheads with most of the ship's officers, sailed with Bering abroad the *St. Peter*. Bering and Chirikov had planned to keep their ships together so that each could help the other in the event of trouble, but a powerful storm set them on separate courses within a fortnight. Although the courses they sailed actually intersected no fewer than twenty-one times during the next four months, neither ship ever sighted the other again.

After the storm, Bering searched for the *St. Paul* for several days and then set an easterly course that carried his ship into the Gulf of Alaska on July 16. Ahead stood Kayak Island, and behind it rose a soaring peak, the highest in North America, that the *St. Peter's* crew named Mt.St. Elias in honor of the saint whose name day it was when they first sighted it. Here was the New World, the chance to extend Russia's power to a third continent and, perhaps, to end the food shortages that continued to plague Siberia's eastern lands. If food could be brought from the North American shores that stood only a few week's sail Kamchatka, Okhotsk and Kamchatka would no longer have to depend upon the fragile lifeline of wagons and barges that still needed several years to carry grain from the Urals to the Pacific. A careful explanation of the new territories that spread out before Bering's ships could yield answers that might affect Russia's position in Siberia for decades to come, or so at least Steller thought. What was needed was time, manpower and the will to explore.

Steller had the will, but neither the manpower nor the time. This was the land he had come halfway around the globe to see, yet Bering at first refused him permission even to go ashore when he sent a company of sailors to Kayak Island for fresh water. After a bitter argument, during which Steller threatened to report to the Admiralty what he called his captain's "intention to force me against my will to inexcusable neglect of my duty", Bering relented on condition that he return the moment that the crew had refilled the ship's casks. "I could not help saying that we had come only for the purpose of bringing American water to Asia," Steller wrote bitterly in his diary. Now that land had been found, Bering thought only of returning to Kamchatka safely and with all speed. "Who knows but that trade winds may arise which may prevent us from returning?" he had said to Steller a day or two before the *St. Peter* reached the coast. "We do not know this country," he warned when Steller begged him to stay longer. "Nor are we provided with supplies for a wintering." These were not the words of an explorer dedicated to discovery. Steller was talking to a man who, having fulfilled his duty in the most precise sense, longed to return home.

While Bering's crew filled water casks on Kayak Island, Steller worked frantically to gather samples of rock and vegetation. In just six hours, he collected 142 plant specimens, including not quite ripe salmon berries, a type of raspberry still unknown in Europe. Thanks to the marksmanship of his single Cossack aide, Steller also brought back a western blue jay, "a single specimen of which I remember to have seen a likeness painted and described in the newest account of the birds and plants of the Carolinas," he later wrote. "This bird proved to me that we were really in America," Steller added. Nothing like it had ever been reported in Europe or Asia.

In a deserted hut that gave every indication of having been inhabited until an hour or so before his arrival, Steller found utensils, a basket, weapons, and bits of food that he hastily scooped up, after leaving an iron kettle, tobacco, a Chinese pipe, and a piece of silk in exchange. Desperate for more time, he sent word to the *St. Peter* that he wanted to go inland, for he had seen smoke on another part of the island, which told him that people were very near. Already his bits of hastily assembled evidence had convinced Steller that the island's people were of the same origin as the Kamchadals of Kamchatka. But he thought that "in view of the great distance we have travelled, it is not credible that the Kamchadals would have been able to get [here] in their miserable craft." The only explanation for their presence, he concluded was that North America stood much nearer to Asia farther north of their landing place. A few more hours of searching Steller thought, would lead him to natives who could verify his conjectures.

Hours might as well have been centuries. When the crew still had twenty casks to fill, Bering suddenly ordered them to come abroad so that the ship could get under way and insisted that Steller return immediately or be left behind. In a rage, the German obeyed. What he said to Bering when he boarded the *St. Peter* we do not know, but he had plenty of vitriol left for the private journal to which he confided his innermost thoughts throughout the voyage. "The only reason why we did not attempt to land on the mainland is a sluggish obstinacy and a dull fear of being attacked by a handful of unarmed and still more timid savages," he wrote. "The time here spent in investigation bears an arithmetical ratio to the time used in fitting out," he added bitterly. "Ten years the preparations for this great undertaking lasted, and ten hours were devoted to the work itself." Once again the broad interests of science had come into conflict with Bering's narrow sense of duty. The Dane's stubborn unconcern for the larger questions of science had cut short the process of discovery at its very threshold.

To be fair, Steller knew at this point that powerful forces were at work in Bering's mind, but for the time being he kept them in his heart and did not mention them in his journal. Struck by the first symptoms of scurvy soon after the voyage had begun, Bering had grown gloomier and troubled as the days had passed. Deep within his inner being, in those dark corners of the soul that men try to wall off from contact with their consciousness, an icy voice was telling Bering that he had not long to live, in those dark corners of the soul that men try to wall off from contact with their consciousness, an icy voice was telling Bering that he had not long to live, even as the shores of North America rose before him. The detachment that separates a dying man from the living world had thus begun to isolate Bering from his officers and crew even before the *St. Peter* reached North America. Uninterested in all and everything around him, Bering seemed to be ruled only by fear that death would overtake him before he returned to Kamchatka. With amazement, Steller noted in his journal that Bering had "even shrugged his shoulders while looking at the land" during their first hours in Alaskan coastal waters. To a man such as he, who had left behind the comforts of his native Central Europe to explore new frontiers of science in an unknown world, Bering's detachment could not be explained. It was as if Columbus had lost interest in the New World the moment his lookout had sighted its shores.

As scurvy weakened Bering's limbs, he remained in his cabin and allowed the reins of command to slip into the unsteady hands of his lieutenants once he had ordered the *St. Peter's* course to be set for Kamchatka. Worried about the scurvy that was spreading among the crew and uncertain about precisely where Kamchatka lay, Bering's officers shifted their headings to produce a cat's cradle of backings and fillings that command forty precious days in sailing from Kayak Island to Alaska's westernmost tip. By the end of August, the *St. Peter* was struggling against contrary winds off the Shumagin Islands, which were normally less than a week's voyage from the point at which they first had sighted land on July 16. By that time, most of the crew were so fiercely at odds with Steller that even when he urged them to search for a spring on one of the Shumagin Islands, they insisted upon filling the ship's casks at shoreside pools with water that inevitably turned brackish. When Steller spoke of collecting plants that could ease the ravages of scurvy, they rejected his advice. The lonely scientist now had to with only the help of Friedrich Plenisner, an obscure German whom Bering had made his draftsman and who, in the course of the voyage, had become his friend.

Steller now had little to work with and less time in which to work. "Our medicine chest," he wrote in his journal, "was filled with enough plasters, ointments, oils, and other surgical remedies for four to five hundred men in case of a battle but had none whatever of the medicines most needed .... against scurvy and asthma." Here Steller's knowledge of botany had put him well ahead of his time. Not until the 1750s would European mariners first discover that fruits and vegetables with a high content of Vitamin C could ward off scurvy. But Steller already knew enough about the disease to think that infusions made from spoonwort could restore movement to its victim's limbs and that dock, one of the most common weeds of the field, could produce a tea that would fix their teeth firmly in their gums once again. He was amazed to find that both grew on the Shumagin Islands-yet the shop's officers, according to his angry journal account, would not assign men to help bring them abroad. Scorned and left to themselves, Steller and Plenisner had time on the Shumagins to gather only the small amount of herbs that could save their lives, prolong Bering's and keep the crew alive for a few weeks longer. Many of the crew would perish in the days ahead for want of the medications Steller had begged them to collect during the last few hours they spent on land before winter.

When the crew of the *St. Peter* set sail for Petropavlovsk from their sheltered anchorage in the Shumagins on September 6, 1741, they began a struggle against wind, waves and scurvy that nearly took them over the brink of disaster. By the third week of October, the ship was wallowing in heavy seas because none of the sailors was strong enough to set the sails. "By now, so many of our people were ill that I had, so to speak, no one to steer the ship," Bering's second in command, Sven Waxell reported, later reported. "When it came to a man's turn at the helm", he went on, "he was dragged to it by two other of the invalids who were still able to walk a little, and set down at the wheel [where]...he had to sit and steer as well as he could[until]...he could sit no more." The weather grew steadily worse. "The winds were violent...to say nothing of the snow, hail and rain," Waxell remembered. "There were ...only three, one of them being the Captain Commander's man, [who] could come on deck," he added. "all others were sick unto death." Too sick to take readings from the sun and stars for several weeks, the crew of the *St. Peter* had no idea of their location when October turned into November. "Our ship was like a piece of dead wood," Waxell wrote. "We had to drift hither and thither at the whim of the winds and waves."

Then, two months after they had left the Shumagins, the *St. Peter's* crew caught sight of land. "It is impossible to describe how great and extraordinary was the joy of everybody at this sight," Steller wrote in his account of that day, November 5. "The half-dead crawled up to see it," he went on, "and all thanked God heartily." The snow-covered mountains that loomed ahead looked so like Kamchatka that the crew at first thought they had come directly to the entrance of Avacha Bay, the gateway to Petropavlovsk harbor. When a few hours more sailing proved them wrong, the officers insisted that, even though the land ahead was not the Avacha headlands, they must be some part of the seven-hundred-mile-long Kamchatka peninsula nonetheless. Once they were on land, Waxell explained, it would be only a matter of sending someone to Nizhnekamchatsk for horses to carry the sick. A council held in Bering's cabin voted to land, although Bering himself begged them to sail on to Avacha Bay. But Bering was too sick to rule his crew any longer. When one experienced seaman urged the men to follow Bering's advice, Steller reported, the others shouted him down with cries of *"Von"* (get out) *"Molchi"* (shut up!).

Led by Waxell and the first mate, the *St. Peter's* officers sent a boat ashore, and when they discovered that they were on an island, they decided to remain there for the winter nevertheless. Since the ground was by that time covered with snow and everyone was too weak to do any heavy work, the men scooped cavelike shelters out of the sand for winter quarters and then stretched scraps of canvas over them to ward off the snow, wind and rain. These became the lazarets for the invalids who were brought ashore a week later. Some of the sickest men died the moment they were taken from the *St. Peter's* hold and the fresh air reached their lungs. Others reached the shore and lay in agony, slowly starving. "As their gums were swollen like a sponge, brown black, grown over the teeth and covering them," Steller noted in his journal account of their first days ashore, "they could not eat anything because of the great pain."

His legs grotesquely swollen and already cold, Bering lingered at the edge of death for three long weeks. Then, on December 8, 1741, he passed into the next world, already half buried, Waxell remembered, because the sand from the sides of his dugout had kept trickling down upon his legs and lower body. He would undoubtedly have remained alive if he had reached Kamchatka and had only had the benefit of a warm room and fresh food," Steller concluded. "As it was," he added, "he perished...from hunger, thirst, cold, hardship and grief." Thirty others of the crew followed Bering in the next month, and it took several months more for the survivors to regain their strength. Only Steller and Plenisner remained unaffected by the ravages that scurvy visited upon the crew, most probably because of the herbs and plants they had gathered during the *St. Peter's* brief stay on the Shumagins.

Now condemned to spend the winter where driftwood was the only fuel and trying to hoard their small reserves of flour and groats in case they remained alive to sail away in the spring the crew of the *St. Peter* had to live on what their island refuge provided. Stewed in oil rendered from the rank carcasses of two dead whales that washed up on shore, sea otters and fur seals, the flesh of the latter describe by Waxell as "revolting because it has a very strong and very unpleasant smell, more or less like that of an old goat," became their main food. Supplemented by potions that Steller prepared to help calm the ravages of scurvy from whatever roots and grasses could be found beneath the snow, this was all the crew had to eat until spring. Some of them continued to linger at the brink of death, while others lived on the verge of madness. For the moment, there was no escape for anyone from the cold, the emptiness and the blue foxes that had no fear of men.

"They attacked the weak and ill to such an extent that men could hardly hold them off," Steller wrote of the blue foxes that besieged their camp day and night. "One night when a sailor on his knees wanted to urinate out of the doorway," he went on, "a fox snapped at the exposed part and, in spite of his cries, did not soon want to let go. "Angry and looking for diversion, the crew tortured the foxes in return. "Inasmuch as they left us no rest by day or night," Steller explained, "we indeed became so embittered against them that we killed young and old, did them all possible harm, and, whenever we could, tortured them most cruelly." "Of some we gouged out the eyes," he explained in his description of the war that Bering's crew waged against the animals. "Others were strung up alive in pairs by their feet," he added, "so that they would bite each other to death." Still, neither vigilance nor cruelty could drive the animals away. Instead, Waxell added, "they are the hands and feet of the corpses before we had time to bury them."

Chance had cast Bering's crew upon the island that now bears his name and that, ironically, stands less than three hundred miles from Avacha Bay. Yet that haven remained a world away for men whose ship had been damaged beyond repair by the winter storms. To build a new ship on an island without trees meant that every plank and spar of the old *St. Peter* had to be put to good use when the crew's forty-six survivors began to break her up under the command of Lieutenant Waxell in the spring. By midsummer, they had finished a vessel barely large enough to hold them all. "Exceedingly anxious for deliverance from this desert island," Steller wrote, they set sail on August 14, 1742. Now the winds and the currents worked in their favor, and in less than three days they caught their first glimpse of Kamchatka. On August 27, 1742, they sailed into Avacha harbor and anchored at Petropavlovsk. It had taken fifteen months to complete a voyage that, under normal circumstances, should have taken less than three. "Surely," Steller concluded in one of his reports," we could be all the more thankful not only for our rescue from the most imminent perils of the sea but also for our preservation." Nonetheless, the scientist in Steller thought the sacrifice well worth the pain. "I would not," he wrote to a friend, "exchange the experience of nature that I acquired on this miserable voyage for a large [sum of] capital."

Including the men who had sailed with Chirikov on the *St. Paul* few others shared Steller's opinion. In keeping with the plan that he and Bering had agreed upon before they left Petropavlovsk, Chirikov had sailed east after their ships had become separated. A full thirty-five and a half hours before Bering sighted Mt.St. Elias, the *St. Paul* reached the North American coast, but farther to the south and east of Prince of Wales Island. Yet Chirikov's crossing of the North Pacific proved to be only moderately less ill-fated than Bering's, and misfortune actually struck the *St. Paul* sooner. When its crew lost both of the ship's boats trying to go ashore for fresh water just two days after they reached the New World, Chirikov's ship faced one of every seaman's worst nightmares. "We had only 45 casks of water," Chirikov wrote in his report to the Admiralty, explaining why he and his officers had decided to return to Kamchatka before they had explored the North American lands that lay before them, "[and] we did not know whether they were full or [having leaked during the voyage] partly empty."

With better navigation and more luck than the *St. Peter*, the *St. Paul* returned to Petropavlovsk ten weeks after it had sighted North America, but that was not soon enough to spare its crew, which had lived mainly on salt meat, from the ravages of scurvy. By mid-September, "some of the men were so feeble that they could not even come on deck," Chirikov reported. A fortnight later, he added, "all members of the crew were down with scurvy...[and]I was so ill and so weak that I expected death at any moment." When the *St. Paul* reached Petropavlovsk, Chirikov had to be carried ashore. De la Croyere, the astronomer and cartographer who had sailed with him that summer, died just before the *St. Paul* dropped its anchor.

Although neither the *St. Peter* nor the *St. Paul* managed to explore the lands that would soon become known as Alaska, the expedition that Bering and Chirikov led into the North Pacific had unexpected and important consequences. Bering had set sail just more than a decade after the Kyakhta Treaty of 1727 had opened a larger market for Russian furs in China than ever before, and at precisely the time when the yields from Kamchatka, the last of the Siberian fur grounds to be invaded by Russian trappers and tribute collectors, had begun to wane. Where were the pelts for Russia's new China trade to come from? And how was the demand from Europe to be met? At this most critical of moments for the Russian fur trade, Bering's expedition found a source of new furs that shifted its center for a century to come.

The Russians had first found sea otters along the coast of Kamchatka at the beginning of the eighteenth century and had immediately dubbed them "Kamchatka beavers." larger than the best Arctic fox and more valuable than the blackest sable, the pelt of a sea otter might contain as much as twelve square feet of precious fur. Its gloss, Steller once explained, "surpasses the blackest velvet," and the silvery white hairs scattered here and there added highlights that no other fur could match. Although sea otter pelts tended to stiffen in severe cold and were thus undesirable for coats or cloaks, the Chinse used them with particularly brilliant effect to trim garments made of other rich materials. At the time of the *St. Peter's* voyage, a single prime sea otter pelt brought forty rubles in Irkutsk and as much as a hundred in Kyakhta. By the end of the century, when the Russian's greed and persistence in hunting sea otters had all but eliminated them from the waters around Kamchatka, the Komandorskie Islands, and the Aleutians, those prices rose by nearly ten times.

Just at the time when the sea otters were beginning to disappear from the coast of Kamchatka, Bering's crew found them in abundance on Bering Island. No one had a moment's doubt that a new treasure had been found, and the men who remained alive when spring came in 1742 used every bit of their returning strength to hunt them down. The tiny ship they built from the wreckage of the *St. Peter* had barely enough room for men and food, yet somehow its crew stuffed in nine hundred precious sea otter pelts as well, knowing that they had found the new "golden fleece" to take the place of the declining numbers of sable, marten and fox on the Siberian mainland. Within a year, the uninhabited islands of the North Pacific began to swarm with men in search of the sea otter. For another century, the Russians would follow this new source of wealth across the North Pacific and through the Aleutians to the lonely northwestern shores of North America.

Chapter-15

The "Russian Columbus"

FUR FROM THE SEA- anew golden fleece that had to be pursued through the icy waters of the North Pacific, stirred a new frenzy in Siberia after the *St.Peter* returned. Only a handful of men could pursue the golden fleece of the North Pacific. To hunt sea otters required a ship and the promise of profits large enough to justify the danger to the men who sailed as her crew. The risk to life and limb was very great. The men who hunted sea otters in the North Pacific often sailed in fragile craft because the iron needed for the spikes and braces of stronger ships could not be had on Siberia's Pacific coast. Their planks "sewn' to rough-hewn keels with willow branches and leather thongs, such ships could often not stand against the force of the winds and the waves. During the second half of the eighteenth century, as many as two out of every five vessels that sailed into the North Pacific in search of furs never returned again.

Sea otters were easy to kill once they were found. "A sea otter prepared itself for death by turning on its side, drawing up its hind feet and covering its eyes with its fore feet," Steller wrote in describing how his scurvy-weakened shipmates killed the defenseless animals on Bering Island with a single cudgel blow on the head. The first hunters found the sea otters such easy prey that they killed almost two hundred thousand of them during the half century after the *St. Peter's* return. They hunted the larger and even more numerous fur seals in their hundreds of thousands. In one massive four-year hunt on the Pribylov islands, the Russians butchered nearly a million fur seals, and then after realizing that their ships could not hold that many pelts, left a third of them to rot on the beach. Because female sea otters had only one or two pups every year, the onslaught of the Russians made their extinction all but certain.

By 1750, expeditions sailing from Okhotsk and Petropavlovsk had wiped out the sea otters that once flourished on the shores of the Kamchatka and the nearby Komandorskie islands and had moved their hunting grounds into the Aleutians. Beginning with the westernmost island of Attu and working east to Unalaska and Unimak, they sailed farther and stayed out longer until only the largest and most elaborately organized expeditions could yield the profits their backers expected. By 1780, the deadly combination of greater risks and shrinking catches had driven the ships of all but seven Russian companies from the North Pacific.

The farther east they went, the more they needed a permanent base. Once they reached the Alaskan mainland that meant fighting the Tlingit Indians. The Tlingit were a warlike people. "They are courageous by nature, and accustomed to enduring every deprivation and physical pain," one officer wrote three quarters of a century after the Russians had built their first settlement in Alaska. "if they were to unite under the leadership of a brave chief," he added, "they would easily conquer our settlements and kill all the Russians." To carry the fur rush into the lands of the Tlingit therefore demanded courage and careful preparation, all the more so since the English seemed likely to challenge any Russian attempt to secure a foothold on the North American continent. Unwilling to risk direct confrontation with British seapower, Russia's empress, Catherine the Great dared not have the Imperial Navy carry her flag to Alaska as her predecessor had hoped Bering would do. That mission now would have to be accomplished by one of the handful of daring

Siberian entrepreneurs who continued to send their ships into the North Pacific.

Perhaps the richest and most adventurous of those who hunted sea otters in the North Pacific, Grigori Ivanovich Shelikhov took up the challenge and staked Russia's claim to Alaska. Celebrated as the "Russian Columbus" by the poet Derzhavin, Shelikhov came from the south-central Russian province of Kursk, which has been described as being "famous for its nightingales and race-horses" but having nothing to do with the sea. Already middle-aged when he arrived in Siberia in the 1770s, he married a well-to-do young widow whom he took with him on his explorations and whose fortune he invested with such profit in the fur trade that he rose into the ranks of Siberia's richest merchants in less than a decade. Shelikhov ships, money or goods played a part in two out of every five fur-hunting expeditions that sailed from Okhotsk and Petropavlovsk during the last quarter of the eighteenth century. To broaden his opportunities still further, Shelikhov joined forces with Ivan Golikov, a countryman from Kursk who shared his vision of raising an empire on a new continent. In just twelve years, the two men shipped almost ninety thousand pelts worth nearly a million and a half rubles from Siberia's Pacific ports to their headquarters in Irkutsk.

In 1783, Shelikhov and Golikov formed a company "to sail to the land of Alaska[Alaska], which is called America, to islands known and unknown, in order to trade in furs, make explorations, and arrange voluntary trade with the natives. Intending to reach North America in time to build quarters before winter, Shelikhov and his wife, Natalia, set sail from Okhotsk late that summer in a new ship they had christened *Three Saints.* Foul weather forced them to spend the winter on Bering Island, and they did not reach Alaska's Kodiak Island until late the next summer. Just before the snows fell, they built the first huts and palisade defenses for Three Saints's Harbor and claimed Alaska for Russia.

Working along with their crew, Shelikhov and his wife spent the next nine months collecting pelts, fortifying their settlements' defenses, and building a second outpost at Kennai Bay, now Cook Inlet. When they set sail on their return voyage with a cargo of precious furs in the late spring of 1786, they knew that settlements in Alaska would not be all that they had hoped, but thought that the profits would be worth the cost nonetheless. How far their effort could go without direct support from St. Petersburg they did not know. That would depend in part upon Empress Catherine the Great's estimate of the risks involved and the measure she took of her potential enemies in Europe.

Extending the Russian's empire to a third continent promised to create major new policy dilemmas for a nation whose rulers felt ill at ease with the sea. It also posed immense problems of supply and communication, for the imperial capital of St. Petersburg stood nearly a world away from the tiny log blockhouse that Shelikhov had raised on the edge of Three Saint's harbor. Between them stretched a fragile lifeline across six and a half thousand miles of Eurasia and two thousand more of the North Pacific. Nothing demonstrated more vividly how tenuous the connection between St. Petersburg and Russia's new eastern outpost could be than the Shelikhov's own experiences in sailing to Alaska and back. Under good conditions, it took less than two months to travel from Irkutsk to Okhotsk and another three to sail to Alaska. Yet it had taken the Shelikhov's nearly a year to reach the North American continent and only a few weeks less to return to Irkutsk after fighting storms at sea and blizzards on land.

Shelikhov returned to Irkutsk hoping (with his partner Golikov's help) to convince Siberia's governor-general, Ivan Yakobi, to put the case in for his grand Alaskan venture before the empress, the Ruling Senate, and the Collegium of Commerce in St. Petersburg. A German nobleman who served Russia's sovereigns as his father had before him, Yakobi commanded the respect that Shelikhov and Golikov could not hope to win at a court whose empress had once spoken of Siberia's merchants as "irresponsible and malignant people", and they knew they could not succeed without his support.

What the two men wanted, Yakobi later explained to the empress's advisors in phrases that fairly dripped with concern for the national interests of the Russian Empire, was soldiers to defend their company's holdings in Alaska, a subsidy from the Imperial Treasury to help finance future ventures, and a monopoly of the Alaskan fur trade. In return, Shelikhov and Golikov would plant Russia's flag firmly on the North American continent.

Far too clever not to see what lay beyond all this, Catherine the Great listened to the case Yakobi presented. She invited Shelikhov and Golikov to St. Petersburg. She gave them honors that cost her nothing but withheld the monopoly they sought. "to reward the zeal of the merchants Shelikhov and Golikov," she announced, "we bestow on each of them a sword and a gold medal to wear about the neck." A special citation from the Senate, she promised would set forth "all their high-minded exploits and activities for the benefit of society," but, she added somewhat regretfully, "at present it will be impossible to provide their company with money or with a military detachment because of the need for troops in Siberia, where there are hardly enough for present demands."

Privately, the empress remarked that the merchant's request for the Imperial Treasury to grant them an interest-free loan for two decades reminded her of a tale she once had heard about a man who promised to teach an elephant to talk in thirty years. Why would it take so much time to educate the elephant? Someone asked. "Either the elephant or I will die," Catherine quoted the would-be educator as saying, "or else the man who gave me the money for the elephant's education will!" She might also have added as reasons for treating the schemes of Shelikhov and Golikov with caution the facts that Spain's hurried movement of its garrisons into Upper California gave her little comfort and that she took no pleasure from the speed with which the British had begun to send their ships to trade for pelts on Vancouver Island and the coast farther north. Catherine knew that the confrontation that was brewing between Spain and Great Britain on the western coast of North America could expand to include Russia all too easily. Already at war with the Ottoman

Empire and Sweden, she had no intention of burdening Russia with more enemies than she dared to fight at once.

In the meantime, the Siberian magnate Pavel Lebedev-Lastochkin demanded a share of the North American fur trade and the handful of Russians in Alaska began to wage war against each other just when they needed to untie against the Tlingit, the British and the Spanish. Soon after Shelikhov died in 1795, these conflicts spread to St. Petersburg. For the better part of a decade, Shelikhov and Golikov had been paying one of the empress's favorites handsomely to keep their competitors from making headway at court, but that line of defense collapsed when Catherine died in 1796 and her favorites fell out of favor. Without a defender at court, the future of the commercial empire the two partners had dreamed of building in North America seemed very dark.

Amid a flood of lawsuits brought by her enemies after her husband's death, Natalia Shelikhova had to face the defection of Golikov, who joined the rival Irkutsk Company, headed by the powerful merchant Nikolai Mylnikov. Shelikhova held her own well enough for a time but she could not fight enemies in Irkutsk, Alaska, and St. Petersburg all at once. She therefore gave greater authority to direct her company's Alaskan affairs to Aleksandr Baranov, the man her husband had chosen as his chief agent in North America. At the same time, she turned to her son-in-law, Privy Counselor Nikolai Petrovich Rezanov, to represent her in St. Petersburg.

Rezanov, whose patent of nobility dated back to the days of Ivan the Terrible, had become the commander of Catherine the Great's personal guards at the age of twenty-three and had managed to stay in the good graces of her son and heir at the same time. Rich, handsome and powerful, his sad eyes and pouting lips making it appear that he had placed his heart ahead of his head. Rezanov had had his choice of any number of eligible young ladies whom hopeful parents pushed his way, but, perhaps, sensing that any marriage in the capital would make new enemies as well as friends, he had looked to the provinces when he decided to take a wife. Charmed by Shelikhov's young and, according to some accounts, "extremely seductive" daughter Anna during a visit he made to Irkutsk on the empress's business in 1793, Rezanov married her before he returned to the capital that fall.

The picture of aristocratic elegance a close friend of the governor-general of St. Petersburg, and a friend of the new emperor Paul, Rezanov had the influence at court to turn his mother-in-law's difficulties into astounding success. Combining his influence in St. Petersburg with Shelikhov ships and capital, Rezanov worked closely with her and Baranov to lay the groundwork for the enterprise that would give them control of the entire Alaskan fur trade. In 1797, he merged the Shelikhov American Company into the famous Russian-American Company, which would dominate Russian ventures on the North American continent until it was liquidated in 1881. With more power than old Shelikhov had ever dreamed of and with its foot firmly set in the New World, the Russian-American Company that Rezanov and Madame Shelikhova founded in 1799 became the main instrument for moving the Eurasian Russian Empire into the continent of North America.

Aleksandr Baranov became the key to realizing Rezanov's grandiose vision in North America. In search of greater prospects and new worlds to conquer, Baranov at the age of thirty-three had left his wife and daughter in a town near Russia's Finnish border and had gone to Siberia in search of a chance to rise above the limited opportunities offered by life in European Russia. For a while, he had combined tax farming with distilling, glassmaking, and fur trading with some success, but his enterprises had eventually turned sour. On the brink of bankruptcy in the late 1780s, he had jumped at Shelikhov's offer to become his chief agent in North America. Like so many who crossed the North Pacific in those days, he had arrived at Three Saint's Harbor after a harrowing voyage.

To the Shelikhov company outpost at Three Saint's Harbor, Baranov brought wisdom, energy, and daring. His wide-set eyes and balding head gave him more then look of a government official than a swaggering explorer, and he focused more upon organizing the company's sparse human and material resources in ways that explorers rarely thought of. While Shelikhov had dreamed of founding a great metropolis on Yakutat Bay that, according to one of his last letters, would be distinguished by "great squares.... on which, in time, obelisks would be raised 'in honor of Russian patriots'," Baranov concentrated more sensibly upon building the better-situated port of Novo-Arkhangelsk (present-day Sirka) on the island that today bears his name. There, where the sun shone less than a third of the year but the winters were so mild that raspberries sometimes bloomed in March and ripened in May, Baranov built the center from which the Russian-American Company would tighten its hold upon Alaska, reach out toward the lands of northern California, and twenty-five years later, make a brief but spectacular lunge toward Hawaii.

Tough, stubborn, and described by Washington Irving (in his memorable tales of *Astoria*) as a "rough, rugged, hard-drinking old Russian [who was] somewhat of a soldier, somewhat of a trader[and] above all, a boon companion of the old roistering school with a strong cross to bear", Baranov labored to strengthen Russian authority on the North American mainland. The Russian-American Company was losing more than half of the ships it sent to Alaska in those days, and hostile Tlingit Indians, spurred on by rum and guns supplied by American and British traders who were trying to drive the Russians out of North America, burned several of the company's most prosperous settlements. Baranov lived on the brink of disaster for most of this twenty-nine years in North America. But his success in dealing with one seemingly impossible crisis after another helped the Russian-American Company to survive for another half century.

Using Novo-Arkhangelsk as a base, Baranov extended the influence of the newly formed Russian-American Company southward at the beginning of the nineteenth century. When the expeditions he sent out in search of food and furs reported that the men of John Jacob

Astor's American Fur Company already held the mouth of the Colombia River, he ordered them to continue on to the lands that the Russians called New Albion, where they built Fort Ross some twenty miles north of Bodega Bay on the coast of northern California. Yet the men sent to farm in northern California preferred to hunt sea otters, and the fertile lands around Fort Ross therefore never yielded enough grain (as Baranov and Rezanov had hoped) to feed the Russians in Alaska, let alone those in Kamchatka and along Siberia's Okhotsk coast. In desperation, Baranov expanded his vision of a Pacific empire to include the Hawaiian Islands, where tobacco, pork, beef, salt, vegetables, and tropical fruits promised rich profits to anyone able to carry them away. Yet Baranov's effort, which began with building a stone fort on the island of Kauai, turned sour very quickly. Less than two years of dealing with the Russians convinced Kauai's friendly king Tomari (who had hoped that his new allies would help him against Hawaii's Kamehameha I) to expel then at the end of 1817 and end their dream of transforming the Hawaiian Islands into what the Russian's chief emissary once called "a Russian West Indies and a second Gibraltar".

Times were changing, even as Baranov tightened Russia's grip upon the lands around Fort Ross and reached toward Hawaii. In the less turbulent days that followed the liberation of Europe from Napoleon by the armies of Russia, Great Britain, and Prussia, men who were more concerned with national defense than about opening new frontiers came to power in Russia, and the less adventurous new directors of the Russian-American Company reflected the same outlook. In 1819, they replaced the roistering old bear who had served them o well for almost three decades with paler men who were less impulsive and more willing to observe the limits imposed on them from St. Petersburg.

By that time, the sea otters had been all but wiped out. For a time, the Russians hunted beaver as a replacement, but in the 1830s silk began to replace beaver in the manufacture of men's hats in the West. Political upheavals in China undermined the demand for furs there. By that time, China's growing weakness also made the Amur valley and the rich oases of Central Asia more accessible targets for Russian colonial ventures, for it had become clear even before the middle of the century that the Chinese could not defend their interests in the lands just beyond Siberia's southern frontier. With its interest in these new directions, the Russian government sold Alaska to the United states in 1867. As Kamchatka and the Pacific shores of Siberia became the easternmost boundaries of the Russian Empire, the tsar's dreams of ruling on three continents came to an end.

During the years when the Russians were pursuing their dream of empire in North America, Siberia had changed from a land of scattered trading posts into a more closely held part of the Russian Empire. Towns and cities were beginning to flourish, trade was prospering, and colonists were beginning to put down deeper roots. Transportation was becoming more certain as roads took shape and distances shrank accordingly. By the end of the eighteenth century (according to a hopeful estimate produced by the Ruling Senate), a government courier could travel the sixty-five hundred miles from St. Petersburg to Okhotsk in less than eighteen weeks, and Nerchinsk could be reached in seventy-five days. A hundred day's journey separated Yakutsk from St. Petersburg, the same document reported, and the towns of western and central Siberia were much closer than that. The Ruling Senate claimed that Omsk, Tomsk, Tyumen, and even Yeniseisk could all be reached from St. Petersburg in sixty days or less. During the century since Peter the great had mounted the throne, it had become possible for Russian sovereigns to make decisions and have their orders carried out-except in Alaska and California-within a matter of months.

Still, travel through the Siberian wilderness remained unpredictable and dangerous. Quagmires that offered neither shelter nor vegetation and ice fields created by permafrost continued to claim the lives of travelers and horses in the wild lands east of Irkutsk even into the nineteenth century. In 1811, it took Captain John Cochrane of England's Royal Navy a full seventy-five days to travel from the Kolyma valley to Okhotsk on a journey that he called one of the most trying of his life. "The difficulties I had to contend with," he wrote to Siberia's governor-general Mikhail Speranskii, after he reached Okhotsk, "surpassed everything of the kind I have seen before and required every exertion of mine to conquer." Nonetheless, Siberia's shrinking distances were making the sum of its separate parts more measurable and accessible. No longer an exotic and isolated frontier land, Siberia now lay within easy reach of anyone who wanted to exploit it in the interest of the government that ruled it from St. Petersburg.

Chapter-16

Siberian Frontier Life

THE COSSACKS, *PROMYSHELNNIKI* AND FOREIGN TRADERS who formed the vanguard of Russia's advance across Siberia lived by their wits in groups of five to a hundred in frontier blockhouses and wintering places across five million square miles of Siberian wilderness. Disaster threatened them at every moment, for a minor illness, a fractured limb, or even an unwary step could bring death in a land where distances greater than the length of the British Isles separated one tiny settlement from another. Very few among these men had crossed the Urals to put down roots. All dreamed of leaving the wilds with a fortune in pelts and ivory. Furs they had come for, and furs they meant to have. Unrestrained by law and ruled by greed once they set foot in the wilderness, they abused Siberia's natives without mercy. All along the frontier, Cossacks demanded tribute for the tsar and gifts for themselves. Robbery and extortion acquired nearly as many furs for *promyshlenniki* and traders as did trade. No matter how willingly offered, no amount of furs or food could satisfy Russia's Siberian frontiersmen for long. Always wanting more than their native victims could provide and brutal in the punishments they inflicted upon the people who could not meet their demands, the first Russians who came to Siberia raised around themselves a wall of native resentment that only hardened their lot.

Anyone with daring and enough cash to buy a few goods to exchange with the natives could enter the Siberian fur trade. Although such small traders collected only what furs could be carried on one or two pack-horses, they accounted for hundreds of thousands of sable, fox and marten pelts before the first flush of

Siberia's fur rush subsided. Only small fragments of data about the furs taken from Siberia have survived, but from them we know, for example, that 436 traders and promyshlenniki on their way back to Russia bought nearly 34,000 sable pelts though Mangazeya during the summer of 1630 and that, in the summer of 1641, they bought 75,0000 sable pelts through the tsar's customs house in Yakutsk. Some estimates have set the value of furs gathered annually by private traders during the seventeenth century at something over a third of a million rubles. At that time, a peasant family of four in Russia earned less than a ruble a year from forty acres of good farmland.

At the other end of the scale from the thousands of small traders and promyshlenniki who braved the dangers of Siberia's taiga to collect a comparative handful of pelts a piece stood the few men who never crossed the Urals but sent agents and hired traders to work for them in the fur frontier. Carrying trading goods by the boat and cartload, such traders stayed in Siberia from two to six years, depending on distance, luck and weather. It usually took a winter and two summers to make the round trip between Russia and the towns and frontier forts of western Siberia, and journeys from Tobolsk to Yakutsk and back could take up to three times that long. A man could complete two such enterprises in a decade, even fewer if he went into Siberia's northeast to spend a winter or two working with local trappers. Because half a dozen journeys to distant parts of the fur frontier could consume the best years of a lifetime, space and time held different meanings for any Russian who ventured into Siberia. That one could move from the vigor of youth to the brink of middle age in the course of two or three journeys from Moscow to the Kolyman lands and back imparted an urgent "here and now" quality to life that contrasted sharply with the large amounts of time that travel in Siberia consumed.

Under the best conditions, conquerors and traders had barely seven months after the spring thaws before the Siberian winter drove them to seek shelter in remote wintering places. Often no more than enlarged peasant huts of roughly squared logs and windows made of mica or translucent fish bladders, these became the centers of winter life in the fur frontier. Mixed with the noxious fumes from a single earthen stove, the stench of crowded bodies turned such shelters into monuments to human's ability to withstand suffocation. They ate, slept and fornicated in a single open room. They fought against the boredom of the long Arctic nights, dressed their raw furs, gambled for entertainment, and killed one another for amusement. In the spring such men set out again. They moved into the hinterland with carts and packhorses. They worked their way up and down the network of tributaries that flowed into Siberia's great rivers, portaging across the land bridges that separated one river system from another. Then they took to their boats once again.

It required ruthless governors to hold these frontiersmen and Cossacks in check. The men sent to govern and defend Siberia were no more upright than the rough-and-ready adventurers they came to rule. Although the laws of Muscovy banned the tsar's officials from pursuing Siberia's "golden fleece" for personal gain, most of them thought of very little else. Almost every man who wielded power in the tsar's name enriched himself by trading in furs or by stealing a portion of those that passed through his hands. Sanctified by centuries of tradition, bribery and theft had been commonplace among old Russia's officials since the days of the Mongols. However, the raw avarice of the men who represented the tsar in Siberia rarely found parallels on the European side of the Urals. Murder, extortion, embezzlement, highway robbery, and selling women for furs and ivory all counted among their crimes. "They were without the fear of God and without feelings of shame," one commentator confessed. "Were it not for the fact that the evidence on this point is uncontradictory," he added solemnly, "one could hardly believe that these men were as low and as depraved as the contemporary literature paints them."

The military governors whom the tsar placed in charge of the regions surrounding Tobolsk, Mangazeya, Yeniseisk, Yakutsk and a dozen other vaguely defined territories offended the laws of God and man more than anyone else. Given unlimited opportunities to enrich themselves at the expense of others, thanks to their monopoly on military power and civil authority in the vast lands over which they ruled, these men lived in opulence. One high official brought thirty-two household retainers, a personal priest, and supplies that included 2500 gallons of wine and spirits to Mangazeya in 1635, and each new military governor who came to Tobolsk typically brought with him at least twice that number of relatives and servants and double that quantity of supplies. By the end of the century, the baggage trains of Siberia's high officials had become so huge that they called to mind the progresses of kings and queens in medieval Europe. Since these men served so far away and ruled over such vast lands, the best that could be done in Moscow to control them was to restrict their terms of office to two years in the hope that a brief appointment would limit their opportunities to steal.

Aside from fur, traffic in females was the most lucrative enterprise among Siberia's Russians. Because women rarely entered Siberia's lonely frontier world of their own accord, native women, deported women, women who had accompanied their husbands and been widowed, and any other women whom fate sent across the Urals all became prime objects for sale and trade. "they traded, gambled, mortgaged, and sold their wives and daughters as if they were chattel," one writer reported after examining the first official reports of such crimes. Promising the women of their choice better food and housing, government officials shared out among themselves females deported to Siberia for offenses ranging from adultery to murder. For women thrown into the raw life of the Siberian frontier, the difference between hunger and survival or between a life of deprivation and one of minimal comfort seemed very large indeed. With the "courtship" rituals of the 1940s and 1950s being remarkable similar to those used three hundred years before, women in Siberia would continue to face those choices into the second half of the twentieth century, when deportations to the frontier finally ended. Even the best born among Siberia's native women faced similar fates. One Russian official reportedly forced the wife of a Siberian prince into his harem. Different sources tell us that others followed his example.

Circumstances drove the more common sisters of native princesses into the embraces of ruder conquerors. The Russians forced Siberian natives to part of their annual tribute in women to be sold in garrisons and wintering places for two or three prime sable pelts apiece. Bought and sold in markets, won at card tables, or traded in barracks and resold after a winter's use without loss of value, these women passed from hand to hand, season after season, never knowing what fate had in store for them from one year to the next. From their passing union with Cossacks, trappers, and exiled criminals came children who combined the blood of their Yakut, Ostyak, Samoyed and Mongol mothers with that of the Russians who fathered them. These became the first element in a long and complex heritage that the Russians would bestow upon the Siberians over the next four centuries and that would end in the 1980 with massive environmental poisoning.

After furs and women, trade in illegal liquor held pride of place as the liveliest enterprise on the Siberian frontier. Out of the way stills and backyard breweries yielded immense profits from the rough-and-ready Russian woodsmen who were ready to drink as much of the moonshine that seared the throat and numbed the brain as Siberia's illegal distillers could produce. Quantity counted most, and alcoholic content mattered more than purity. Like their modern-day descendants, who have been known to produce moonshine from almost any organic material known to man, the Russians in seventeenth century Siberia consumed any quantity of any spirit produced at any time from any substance. In vain the authorities condemned producers, sellers and consumers of Siberian frontier spirits to public floggings, fines and imprisonment. Nothing could stop the trade in bootleg spirits from following the Russians across the Urals and nothing could eradicate it once it had taken root.

Even in a crime-ridden land, crime had a price, especially for those without influence or the means to buy it in high places. The tsar's officials sentenced those who could not protect themselves to flogging, mutilation, imprisonment, hanging and an astounding variety of cruel tortures. In a petition to the Siberian Office, a group of insubordinate Cossacks once complained that the military governor of Yakutsk "burned us over a fire, pulled out our navels and muscles with hot tongs." Others reported that the same high official fried men on large iron pans, drove needles under their finger- and toenails, and scattered burning coals onto their bare backs. Criminals had noses, ears and limbs cut off, tongues torn out, and cheeks, foreheads and backs branded. Men sentenced to capital punishment actually fared better than many who were not, for the authorities allowed them to seek new sources of tribute for the tsar in the lands beyond the frontier instead of facing the executioner. At some point in their lives, several of Siberia's most famous seventeenth century explorers-including it seems, Khabarov-went in search of new tribute rather than face the tsar's hangman. For those guilty of lesser crimes, the old Russian proverb that "God is high above and the tsar far away" took on new meaning in a land where it took over two years for a petition for clemency to reach Moscow, he acted upon, and be sent back to Siberia.

Even more than the rough-and-ready frontiersmen whom the tsar's officials tried to intimidate by torture and punishment, Siberia's natives suffered at the Russian's hands. Although the Siberian Office in Moscow warned them not to "insult" them, Siberia's governors-general frequently took chiefs and tribal shamans hostage, kept them in irons, and fed them with the rotten meat kept for sled dogs in the winter. Reports from western Siberia complained that the tsar's officials tortured natives as a matter of course. The military governor of Yakutsk once hanged twenty-three chiefs (for which he received a reprimand), and an official in Okhotsk was reported to have abducted all the children from the local Tungus so that he could sell them back to their parents for a sable pelt apiece.

Disputes about fur tribute usually lay behind these misdeeds. Although the Russians took part of the payment in native women, Siberians had to pay most of the tribute owed to their new masters in furs. Depending upon the time and place, each native male paid one to ten prime sable pelts each year, and extra furs were demanded as "gifts" for the officials who collected the tribute. Siberians who lived within reach of a town or frontier blockhouse, of which the

Russians had built twenty-nine in western Siberia by 1640, took their tribute to the chief official in residence at the end of the trapping season, while special expeditions of Cossacks and tribute collectors made the rounds of those who lived far away. Then the Russians graded the furs, packing fox pelts in bundles of forty, and minks and squirrels in bundles of sixty and eighty before they shipped them to Moscow, where the Siberian Office calculated their value (noting the particular flaws and fine points of each), before storing them away or using them in trade.

Combined with the tax of one pelt out of every ten levied on private traders and *promyshlenniki*, the tsar's fur tribute amounted to over a hundred thousand pelts a year by the end of the seventeenth century. Furs paid the expenses of the tsar's court and the costs of his government and served as gifts to foreign sovereigns and their ambassadors. This was the "soft gold" that helped the Russians to recover from the time of troubles that had preceded the Romanov's rise to power at the beginning of the seventeenth century and paid the cost of winning back the western lands that the Swedes and Poles had seized. Profits from the fur trade would also help to finance Peter the Great's massive campaign to modernize Russia at the beginning of the eighteenth century. In later times, Siberian gold, silver, charcoal and iron ore would help the Russian Empire to enter the Industrial Revolution. In the twentieth century, gold, diamonds, coal, oil, natural gas and timber from Siberia would help to pay the costs of the massive experiment in social engineering that Stalin and his successors imposed upon the Soviet people.

Just as the men who vied for wealth in the great gold rushes of the New World took up other occupations when their gold fever abated, so the Russians in Siberia turned to other sources of livelihood as the fur frontier moved east. At the peak of the fur rush, Cossacks, traders, and *promyshlenniki* had dominated the Siberian frontier, where they lived by violence and were ruled by it. But because the fur frontier advanced so rapidly, the men and women who settled in western Siberia in the 1640s and 1650s no longer had furs as their chief concerns, nor were they foremost in the minds of those who moved into the lands of the Yenisei and the Lena in the 1670s and 1680s. As they put down roots, built homes and businesses, and set the stage for a new way of life, these first *sibiriaki* -Russians who had left their old lives on the European side of the Urals and began to think themselves as Siberians-became the creators of a new Siberian culture. Eventually, some of them would become powerful voices for separating Siberia from the Russian Empire.

Among the men and women who came to build new lives, carpenters, blacksmiths, gunsmiths, boat builders, and scores of other artisans all found a large demand for their skills. Once the fur frontier had moved east, trade was no longer a matter of exchanging knives, kettles, beads, and trinkets for pelts, and by the 1650s, Russian artisans in western Siberia were producing silver, leather, and iron goods to supply a growing demand. By the end of the century, commerce with the lands of Central Asia had added new dimensions to Siberia's economic life, and the growing market for Russian furs in China made the forces shaping Siberian trade more complex. By the 1670s, furs from the frontier lands of North America were beginning to compete with Russian furs in Europe. With their monopoly of the European fur trade being challenged, the Russians saw the markets offered by China and the khanates of Central Asia as particularly attractive new prospects. It was no accident that the tsar began to send diplomats to China during the second half of the seventeenth century nor that working out a trade treaty with the Chinese became an ongoing preoccupation for the Russians from the 1650s to the end of the 1720s.

Chapter-17

Siberian Towns

TOBOLSK EMERGED AS THE CHIEF town of Siberia. The center of Russian government, religion and commerce east of the Urals, by the seventeenth century it boasted a monastery, nunnery and a seminary. Bukharan merchants and Tatar traders brought goods from the East to its markets. The prosperity of its artisans added color and vibrance to its economic life. Almost all of Russia's Siberian trade passed through the hands of Tobolsk merchants. By the end of the seventeenth century, Tobolsk had the largest merchant arcade in Siberia and one of the most prosperous anywhere in Russia. However, Tobolsk was by no means a lonely island in the Siberian wilderness. By the second half of the seventeenth century, Tomsk had begun to play a similar role in south-central Siberia. Yeniseisk, the funnel for all trade passing to and from Yakutsk and Siberia's far east, did the same farther north. East of Yeniseisk, Yakutsk and Irkutsk became the key towns. Although Yakutsk was founded a full quarter century before Irkutsk, eventually the latter became more important. Furs, goods and grain going to and from the lands Dezhnev and Stadukhin had opened in Siberia's northeast passed through Yakutsk. For this reason, it ruled all the lands beyond the Lena during the middle third of the seventeenth century. Then Irkutsk, an outpost founded at the intersection of the Angara and Irkut rivers in 1661, began to collect tribute from the Buryat natives and take control of the Transbaikal lands. Within easy reach of the Chinese frontier, Irkutsk became the link Moscow needed to trade with Peking. Moreover, it was one of those rare regions in eastern Siberia which could grow its own grain. Unlike in Yakutsk and Yeniseisk, hunger was rare in seventeenth century Irkutsk. Its people never depended on grain imports from across the Urals.

In Siberia's western lands, dependence on grain from European Russia was diminishing by the second half of the seventeenth century. More Russians were turning from trapping to farming. Grants of livestock, seeds, tools and cash subsidies from the Royal Treasury were drawing more peasants to Siberia from those parts of European Russia were the soul was poor and farming hard. In Siberia, a family which tilled one acre for the tsar could till five acres of government land for themselves. They escaped the poverty that crushed them back home in European Russia. Some of those who began new lives in Siberia were government-sponsored settlers. Others were runaway serfs. These serfs knew their masters had little chance of finding them once they crossed the Urals.

The newly built town of Kyakhta that stood on Siberia's Chinese frontier became the center of the regenerated East-West trade that had flourished in the days of the Silk Route. A barren spot of ground with neither trees nor shelter in the days of Peter the Great. Kyakhta became one of the greatest trading centers of the Russian Empire within a century. A town built in the midst of nowhere, Kyakhta was the creation of Count Sava Lukich Vladislavich-Raguzhinskii, a Herzegovinian nobleman. Vladislavich had undertaken a number of delicate missions of state for Peter the Great. Many of them were under less than pleasant circumstances. Now, ambassador extraordinary and minister plenipotentiary to Peter's widow, Catherine I, he was to negotiate a treaty to guarantee the borders between China and Russia and open the Chinese frontier to Russian trade. Vladislavich reached Irkutsk in the dead of winter. This marked the first stage of an odyssey across the Mongol steppes and the wastes of the Gobi Desert to Peking. There, the Chinese emperor told him that his ambassadors would discuss trade between Russia and China in the Kyakhta valley. This was a little more than four hundred miles from Vladislavich starting point at Irkutsk.

The next fall, soldiers from Siberia's Tobolsk Regiment laid the foundations for the town of Kyakhta directly across the frontier from the Chinese border town of Mai-mai-cheng. Located 4,329 miles east of St. Petersburg and 1,021 miles north and slightly west of Peking, this was one of the most wretched sites for a town that could be imagined. This was a desolate wilderness. The nearest source of drinking water stood a good half hour away. Firewood had to be carried for at least twenty miles on the backs of men and beasts. "The soil is so poor," England's Captain John Cochrane wrote when he visited the town ninety years after its founding "that even common vegetables are with difficulty raised." A decade after Kyakhta's sesquicentennial had passed, the American journalist George Kennan still found it hard to think of Kyakhta as "the most important commercial point in Eastern Siberia." Why, one Russian asked, "had two powers like Russia and China not treated themselves to something grander?" The answer remained forever a mystery, perhaps known on the Russian side only to the sphinxlike Vladislavich. Outwardly, neither Kyakhta nor its Chinese counterpart Mai-mai-cheng would reflect the wealth that flowed through them in the years to come.

China's Mai-mai-cheng was only slightly less primitive than the Russian's Kyakhta. Just a few hundred yards to the south, Mai-mai-cheng (meaning, literally, "buy-sell city" in Chinese) was more carefully laid out than its Russian counterpart. It had larger houses, cleaner streets and a better water supply. It remained a frontier outpost nonetheless. Here, there were no women, a fact that Captain Cochrane blamed on its resident's preference for pederasty, "that dreadful degeneracy," he wrote," which is said to pervade all ranks of society." More probably, China's emperors banned women from Mai-mai-cheng to discourage merchants from settling permanently in a place where they ran a greater risk of becoming Westernized. For, if the Russians saw Mai-mai-cheng as an outpost of China on the Siberian frontier, the Chinese regarded it as the gateway through which Western culture could cross their north-west border.

In Kyakhta and Mai-mai-cheng, the Russians traded furs, skins, leather, ginseng, and coarse cloth for fine Chinese cottons, dried rhubarb root, silks, and tea for the next two hundred years. At the top of the list of goods coming through the Kyakhta gateway stood Chinese cottons, more than a million yards of which passed into Siberia every year during the eighteenth century. Silk had its place too, for the foulards, brocades, gauzes and crepes that made the journey through Kyakhta held honored places in the lives of eighteenth century Russians and Europeans. Seamstresses and tailors from Moscow to London dressed their best customers in silken satins and velvets, but silk never challenged the position of cotton at the head of Kyakhta's list of Chinese imports. Tea, however, did.

Although often thought of aside from vodka as the Russian national beverage, tea did not become part of upper-class Russian life until the 1770s and 1780s. Before that, it was consumed more in Siberia than in Russia and was brought by the merchants of Kyakhta mainly in the form of hand-packed bricks, which the Siberians infused with mutton fat, salt and rye meal. Durable, easily stored, and weighing about three pounds apiece, these bricks even served for a time as units of currency in specie-starved Kyakhta. Until the 1780s, the three hundred tons of brick tea that entered Siberia through Kyakhta each year far outweighed the more expensive, higher quality loose leaf tea that Russian merchants and nobles were learning to consume. Then at the end of the century, the quantity of loose tea being shipped through Kyakhta soared to almost 700 tons a year. This surpassed the 567 tons of brick tea that Siberians continued to purchase for their own use and for sale to the lower classes in Russia. By then, the value of Russia's yearly tea imports from China stood at nearly two million rubles.

Dried Chinese rhubarb root, from which medicinal infusions could be brewed, was regarded as particularly important in the West. It therefore became a prized commodity in the Kyakhta trade. Both a cathartic and an astringent, rhubarb differed from other purgatives "in that it has a high tannic content thereby preventing free movement of the bowels after complete evacuation has been obtained." Eighteenth century European physicists also prescribed it as a cure for maladies ranging from liver disease to gonorrhea and intestinal worms. They thought it grew best in the highlands of Tibet, Kansu and Chinese Central Asia. Rhubarb was sought after to purge the heaving bowels of middle and upper class patients in every city west of Moscow. The supply of dried Rhubarb coming across the Chinese frontier almost never satisfied the demand. Twenty-five tons of high-quality dried roots passed through Kyakhta every year. There were times when rhubarb sold in St. Petersburg for fifteen times what it cost in Kyakhta. Merchants who bought it at those prices had little trouble selling it in Paris and London at a hefty profit.

For much of the eighteenth century, the Russians used Siberia's furs to pay for the Chinese goods that passed through Kyakhta. In China, high-quality pelts brought up to ten times what they cost nearer to their source. Warm, durable and inexpensive, Siberian squirrel headed the list of Chinese imports. The merchants of Mai-mai-cheng bought more than seven million pelts a year from the Russians toward the end of the century. Ermine, muskrat, sable, fox, beaver, domestic cat, ferret and rabbit accounted for over a million more pelts in some years. By the beginning of the nineteenth century, the Manchu's appetite for furs had outstripped the combined resources of Siberia and the newly discovered Aleutian and Alaskan fur grounds. The Russians began to import tens of thousands of beaver pelts from Canada and the American Northwest to sell in Mai-mai-cheng. Once the gateway at Kyakhta had been opened, the mandarins and merchants of China made Siberia's furs the Russian's true golden fleece. Nothing the Russians sold in the eighteenth century equaled their value, not even the Demidov's iron.

The trade that flowed to and from Kyakhta during the eighteenth century helped the Siberian towns along its route to flourish too. Irkutsk was founded in 1661 as a base for collecting the tsar's fur tribute from the local Buryat natives. It became the chief distribution point for all goods moving to and from Kyakhta. Irkutsk became one of the most rapidly developing towns in Siberia as a result. Here in the Siberian wilderness, 3900 miles from St. Petersburg and farther east than Singapore, Russia's trade with China produced a prosperity that few urban centers farther west could match. Profits from the Kyakhta trade built a stone cathedral, several wooden churches, half a dozen arcades for wholesale and retail trade, ten taverns, a brewery and public baths. Irkutsk boasted a host of government buildings and nearly a thousand privately owned shops and dwellings before Catherine the Great ascended the throne in 1762. During the next forty years, the economy of Irkutsk grew even stronger. More and more merchants and craftsmen settled there and accumulated greater wealth.

The Shelikhov-Golikov enterprises had their headquarters in Irkutsk. So did the firms of Mylnikov and Lebedev-Lastochkin, their chief rivals in the Aleutian and North American fur grounds. The great Irkutsk commercial fair between mid-November and the end of December and gain from mid-March to the end of April, brought together merchandise from Europe, Russia and China. The commodities traded grew more varied as each decade passed. By the beginning of the 1780s, well over three hundred different kinds of Russian goods valued at more than two million rubles changed hands at the Irkutsk fair during the so-called "December market." Even more appeared in the spring. Then, the Shelikhov-Golikov and Lebedev- Lastochkin companies began to pour furs from the Aleutians and Alaska through the town's arcades toward the end of the 1780s. A decade after that, the value of Chinese and Russian goods moving back and forth through Irkutsk passed seven million rubles a year.

Along with Tobolsk and Tomsk, Irkutsk became one of Siberia's leading urban centers. Its population passed the fifteen thousand mark around the beginning of the nineteenth century. The first public library in Siberia opened in Irkutsk in 1782. This was due to a donation of more than thirteen hundred books from the Imperial Academy of Sciences. Before the end of the century, Irkutsk had an amateur theater, a forty-piece orchestra (brought in by Governor - General Yakobi in 1780s), and one of the first centers for smallpox vaccination anywhere in the world. On orders of Catherine the Great, more than fifteen thousand people in and around Irkutsk, including many of the Tungus and Buryats became among the first in the world to be inoculated against smallpox. Buryats, merchants from Moscow, St. Petersburg and Kazan, and a host of lesser Russian towns, Swedes, Germans, an occasional Englishman, an American or two, and traders from many other parts of the world all found places in Irkutsk in those days. One of the emperor's own counselors spoke of the "lively" society and "polished manners' he had found in Irkutsk around 1820. Captain Cochrane who passed through the same year, thought that "its resources would be sufficient even for a capital of an independent kingdom."

Despite its illustrious origins, Yeniseisk had fallen behind Irkutsk by the middle of the eighteenth century. Still, it remained the gateway to the Turukhansk region and the rich fur lands of the central Siberian north. It was the site (at the town of the same name) of the fair at which buyers and sellers of the best pelts of the Arctic gathered in June. In August, on the heels of the Turukhansk fair, came Yeniseisk's own. Here the natural products of Siberia competed with the work of skilled Yeniseisk craftsmen for the attention of merchants from all over European Russia. Goods bought at the fairs of Turukhansk and Yeniseisk was hauled in barges up the Angara to Irkutsk and sent overland to Kyakhta. Alternatively, it was shipped sixty miles overland to Makarskoi Fort, from which it travelled on barges down the Ob and then was hauled up the Irtysh to Tobolsk. Not until the 1760s, when Siberia's Governor General Chicherin built a road across the Baraba Steppe, did Yeniseisk begin to lose its importance as a key transit point between Irkutsk and Tobolsk. Then the line of Siberia's east-west trade moved south, following the completed sections of the Great Siberian *trakt* that connected Irkutsk with Tobolsk.

Situated halfway between St. Petersburg and Irkutsk, Tobolsk stood in the midst of lands in which farming flourished. "It produces a great amount of grain and all the necessities of life in great plenty," one visitor wrote at the beginning of the nineteenth century. "Vegetables of all kinds," he added "grow here in a remarkably fine state, as well as pumpkins, melons and cucumbers." Tobolsk in those days had about the same population as Irkutsk. It was thought to be more cosmopolitan because it stood closer to European Russia. One of the emperor's advisors thought that Tobolsk around 1820 could "fairly stand a competition with that of some of the best provincial towns of Russia." In comparison to other parts of Siberia, Tobolsk could lay a fair claim to enlightenment and culture. The fur trade provided a major source of wealth there too. Tobolsk in 1790 boasted a dozen icon painters, a clockmaker, eighteen silversmiths, forty-five blacksmiths, thirty-five gunsmiths, and a host of tailors, dressmakers, bootmakers, and legions of other artisans.

In 1775, Catherine the Great divided the "Tsardom of Siberia" into two parts. Irkutsk became, officially, the eastern counterpart of

Tobolsk. Between the two "capitals" lay the Baraba Steppe, the swamps and peat bogs of which formed a thousand-mile barrier of land so inhospitable that no one had yet settled within its boundaries. The Baraba Steppe had no villages, no land dry enough to grow grain, little firm ground, and almost no fresh water. Any goods or people being sent farther east had to be hauled in barges against the current along the route formed by the Ob and Angara rivers, even though the distance by water was more than twice as far as by land. To lay a road across the steppe, drain its swamps, and fill its bogs, the government used serf and convict labor upon whose lives it placed little value. "Exhaustion, brutal punishments, fever, typhus, scurvy, and the scourge of malignant anthrax took the lives of thousands," one writer wrote of these labor gangs. "Then new thongs fated for similar deaths took their place."

Still far from being the sort of highway to which travelers had grown accustomed in the West, the Great Siberian *trakt* became Siberia's major east-west artery starting in the 1780s. Mud flowed axle deep in spring and fall, and clouds of dust choked every living thing that moved along it in summer, when many of the streams and ponds that supplied its water turned brackish and foul in the heat. With its closely spaced post stations, the Great Siberian *trakt* was most easily traveled in winter, when snow covered the dust and frost solidified the mud. Then goods began to move in quantities and at a speed not often seen in the West. In the 1790s, when Siberia had become a major link in the trade networks that spanned Eurasia, as many as ten thousand sledges gathered in Irkutsk to carry freight west from Kyakhta once the ground had frozen.

Improvements in transportation narrowed the gap that had separated East and West since Yermak's conquest. Certainly, Siberia's west continued to be better developed than its more distant eastern lands, but the space between the two was closing as the eighteenth century came to a close. Some commentators still complained of the "bourgeois aspirations of the merchant class, the despotic monopoly of the officials[and]the dark ignorance and passive servility of the masses.". The fact remained that there was no titled aristocracy in Siberia. Its merchants became the bearers of culture and civilization. Men and women like the Shelikhovs performed that mission more capably than the Empress Catherine and her courtiers. The latter thought the Enlightenment the exclusive domain of Russia's aristocrats. Even in the 1780s, Irkutsk had its small circle of literati, who discussed the works of Russia's leading writers. Foreign tutors appeared in Irkutsk and Tobolsk despite one governor's complaint that there was "almost no education[there]at all." By the early nineteenth century, one visitor was amazed to see in Irkutsk "even the wines and other luxuries of Europe.... sold at very moderate prices." "The traveler is truly surprised," he added, "not only to find a well-built populous town, but also a genteel pleasant society, and almost all the luxuries of life, in the very heart of Siberia."

Greater opportunities and more abundant food supplies in all but Siberia's easternmost maritime lands began to bring more Russians into the lands beyond Tobolsk as the eighteenth century moved toward the end. In 1700, more than six out of every ten Russians living east of the Urals had settled in Siberia's western Lands. A hundred years later, six out of every ten lived in its central and eastern parts. Between 709 and 1797, The number of Russians in Siberia more than doubled, while the number of natives increased by scarcely two thirds. More people, more trade, and more goods meant that Siberia needed better government. No longer could Siberia's Russians concentrate their energies upon exploiting native fur gatherers. Now they had to govern the vast colonial land that all of them recognized to be the largest part of the Russian Empire.

Chapter-18

Siberia at the Threshold of the Modern Age

HOW COULD SIBERIA BE HELD FIRMLY WITHIN Russia's imperial framework? How could its resources be developed and utilized? How could its natives be transformed from exploited victims into useful citizens? And, most of all, how could the gross abuses of authority and the raw corruption that had been the chief characteristics of Siberia's governors for more than two hundred years be brought to an end? Siberia's seventeenth century military governors had been crude and cruel men who had wielded raw power and abused it with impunity. Their eighteenth century counterparts-the governors appointed by Peter the Great and the empresses Anna, Elizabeth, and Catherine the Great-had reflected the veneer of European civility with which the Russians had begun to overlay their Muscovite heritage, but they had been no less corrupt than their predecessors and no more willing to put the interests of the land and people they ruled ahead of their own. Prince Marvel Gagarin, a scion of one of Russia's most illustrious noble families, had been hanged for all he had stolen during his term as Siberia's governor-general in the days of Peter the Great. Although few eighteenth century Siberian governors combined their prince's greed and daring, hey all claimed an illicit share in Siberia's wealth.

Such men fed upon the poor and weak as well as the strong and prosperous. Even after Governor-General Zholobov, one of Gagarin's successors, amassed a fortune in gold and precious furs, he continued to be petty enough to demand a piece of meat or a few eggs as a bribe from any victim who had nothing better to offer. A squadron of soldiers had to be sent from St. Petersburg to put Zholobov under arrest, yet his successors were only marginally better. The first inspector-general sent to investigate some of the most blatant of Siberia's official's crimes himself extorted 150,000 rubles from the citizens of Irkutsk during his visit, and one of the senators whom Alexander I assigned to wipe out corruption in Siberia proceeded to set himself up in Irkutsk as an Oriental potentate to whom, one chronicler wrote, everyone "bowed low and kept silent." After being raised to the rank of governor-general, this senator left his wife behind in Tobolsk and settled in Irkutsk with a French mistress. Thievery, bribery and graft flourished at every level of Siberia's government, and having unlimited power to get at Siberia's great natural wealth corrupted almost every official who wielded it.

As Siberia entered the nineteenth century, its governors seemed every bit as self-serving as those who had wielded power in the days of Peter the Great, when the main difference between the tsar's officials (to quote one of them) was that some of them stole from the people they ruled "on a bigger scale and in a more conspicuous manner than others." Even when some of these men began to think of governing Siberia more efficiently, they could not set corruption aside. Madame Treskin, the wife of the governor of Irkutsk during the second decade of the nineteenth century, was even reputed to have set up a special store to sell the bribes that petitioners showered upon her, while a much less exalted official who served her husband was found to have acquired furs and other goods worth more than a hundred thousand rubles by the time he was dismissed from office.

Could such a government be transformed into one that put the interests of Siberia and its people ahead of its governors? That was the task Emperor Alexander I set for Privy Counselor Mikhail Mikhailovich Speranskii, his closest adviser on Russian domestic affairs in the years before Napoleon invaded Russia. Alexander named him Siberia's governor-general in 1819, and Speranskii, who had advised the emperor about domestic reforms at the beginning of his reign and later would become famous for codifying Russia's laws, found in Siberia his greatest challenge. There, in a world where everyone violated the law, he had to win respect for the law from governors and governed alike.

Russia had expanded her frontiers a hundredfold from the time of Yermak's conquest. Her domains reached the Pacific Ocean. The period from 1500-1800 constituted a different period of Siberia's history, one of exploration and garnering land. The men involved were frontiersmen, trappers, Cossacks, runaway serfs. Siberia promised freedom and hope. It promised vast empty spaces ready to cultivated. It offered a freer life. It saved people from the tyranny of hierarchies. Unfortunately, this dream of freedom was to vanish in the nineteenth and twentieth centuries. Siberia came to represent one vast prison, one infamous abyss which represented the death of human rights, the absolute degradation of human beings. It represented cruelty, inhuman behavior, forced labor and all manner of unjust actions on the part of governments.

This book is written specifically only about the earlier period, before this degradation of human values took place. Before the virgin forests of the taiga were denuded and before Siberia was turned into an industrial garbage dump polluting the atmosphere with millions of tons of chemical effluents. It is written with the aim to remind one that a vast portion of the Asian continent became part of a European country and that this important event in the history and geography of nations has gone unnoticed.

Maps

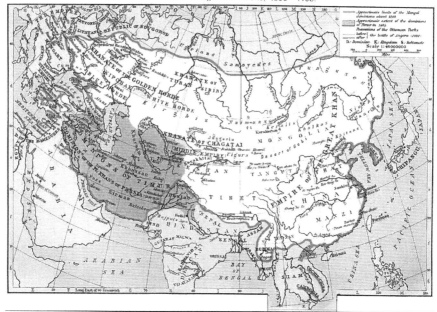

The Mongol Empire

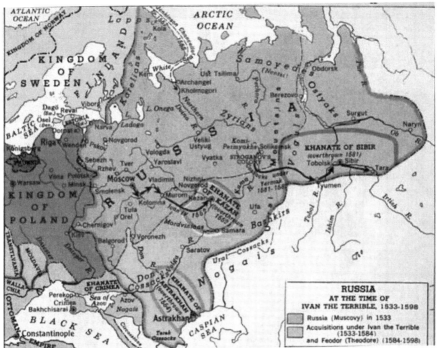
Russia at the time of Ivan the Terrible

Russian Acquisitions and Expansion

Conquest of Siberia

Siberian River Routes

Antique Map of Lake Baikal

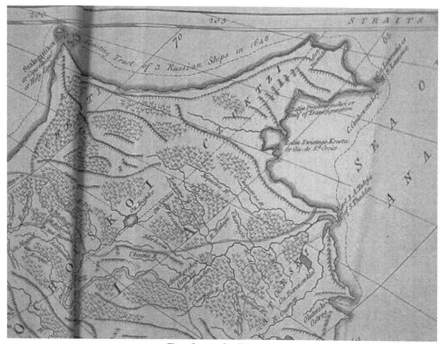
Dezhnev's Route

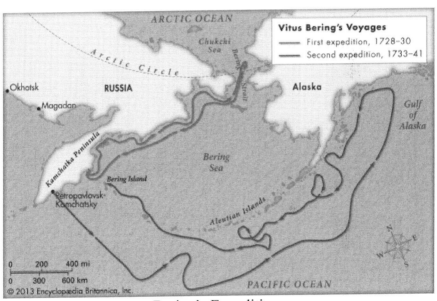
Bering's Expeditions

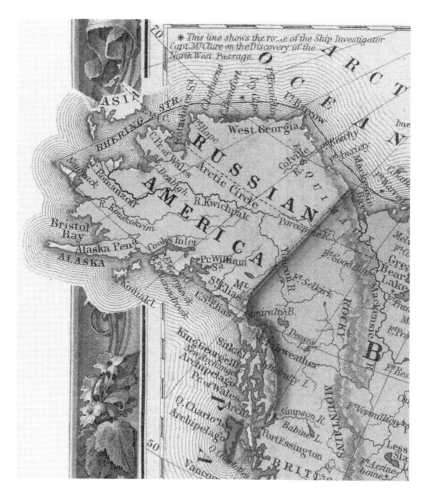

Russian America

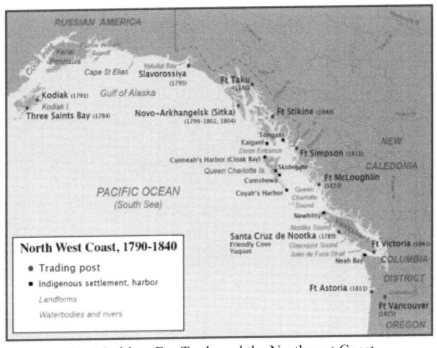

The Maritime Fur Trade and the Northwest Coast

Illustrations

Chinghiz Khan

Batu

Ivan the Terrible

Storming of Kazan

Yermak Timofeyevich

Yermak's Conquest of Siberia, a painting by Vasily Surikov

Laminar armor from hardened leather enforced by wood and bones worn by Chukchi, Aleut, and Chugach (Alutiiq)

Koryak Armor

Tara Gate in Omsk city, formerly a part of the Omsk fortress

The tower of the 17th-century Russian Ilimsky ostrog, now in Taltsy Museum in Irkutsk, Siberia.

The 17th-century tower of Yakutsk fort

A 17th-century koch in a museum in Krasnoyarsk. Kochs were the earliest icebreakers and were widely used by Russian people in the Arctic and on Siberian rivers.

Siberian peoples as depicted in the 17th century Remezov Chronicle

Printed in Great Britain
by Amazon